Praise for *The Miracle Workers o...*

"For years' people have been touting the healing effects of hyperbaric oxygen therapy. However, anecdotal data does not win political support or funding, but with the publication of *The Miracle Workers of South Boulder Road*, pragmatic and practical HBOT data is now available. With a medically approved clinical plan utilizing hyperbaric oxygen therapy and psychological counseling, The Rocky Mountain Hyperbaric Institute has evidence to prove that they have healed more than 300 US combat veterans of TBI and PTSD symptoms and have allowed them to return to normal lives with their families, careers and education. Now is the time for Congress, VA and FDA to recognize that this treatment has the power to heal and cure our nation's veterans of the signature injury of this current war."

~ Ralph P. Bozella, Vietnam Combat Veteran,
U.S. Army Chairman, National Veterans Affairs
& Rehabilitation Commission,
The American Legion Past President,
United Veterans Committee of Colorado

"At long last, something that really heals—Hyperbaric Oxygen Therapy! Thank God for Eddy and Ryan and their courage and determination in establishing RMHI for veterans and others. And many thanks to Colonel Bob Fischer and Grady Birdsong, both Marine Vietnam combat veterans, for revealing the best kept secret for curing TBI and PTSD in this great book."

~ Timothy C. Hall, former Deputy Executive Director
Colorado Department of Human Services
Veterans and Disability Services Division
Denver, Colorado
Member of United Veterans Committee of Colorado

miracles. That is not the true miracle, however. These Miracle Workers are applying the science of medicine through the genius and guile of people who truly desire to serve their fellow mankind. No, the true miracle is their courage and determination to fight for these people and deliver hope and help! *The Miracle Workers of Boulder Road* inspires me to continue my advocacy to battle for our Veterans who suffer the debilitating effects of TBI-PTSD. Drugs from the VA don't work! Families continue to be stressed, tattered, and broken from these wounds—the treatment is available, yet so elusive, as we struggle with political pushback, naysayers, and bureaucrats. I advocate for hyperbaric oxygen therapy because of a family at my church. Their son, a Veteran of Iraq, committed suicide—neither he nor they ever knew about HBOT. The drugs he was prescribed didn't work—for him, nothing did. I see his parents at church often and I think of the 'Miracle Workers.' I wish I would have known and been able to help sooner."

~ Brigadier General James L. Bauerle, USA (Ret)
Veterans Advocate
The Military/Veterans Coalition of Indiana

The Miracle Workers of South Boulder Road

HEALING THE SIGNATURE WOUNDS OF WAR

RYAN FULLMER   EDDIE GOMEZ

Authors

Bob Fischer & Grady Birdsong

Foreword by Colonel James K. Wright, M.D., USAF (Ret)

Afterword by Xavier Figueroa, Ph.D.

BIRDQUILL LLC

The Miracle Workers of South Boulder Road:
Healing the Signature Wounds of War
by Robert L. Fischer, Grady T. Birdsong

Published by

BIRDQUILL LLC
Denver, CO

Contact the authors/publishers at
Bob06usmc@gmail.com
GradyTBirdsong@aol.com

ISBN: 978-0-9976068-0-5 (print)
SBN: 978-0-9976068-1-2 (e-Book)
Library of Congress Control Number: 2016907988

Cover and Book Design by Nick Zelinger, www.NZGraphics.com
Editing by Donna Mazzitelli, www.MerryDissonancePress.com

1. MEDICAL/Neuroscience
2. MEDICAL/Neurology
3. PSYCHOLOGY / Psychopathology / Post-Traumatic Stress Disorder (PTSD)

First Edition

Printed in the United States of America

To my wife, Karen, for your continuing support and encouragement during the writing of The Miracle Workers—*your reading and critiques helped me clarify and complete some of the more difficult passages. You have also been there for me during* Covan, Guerrilla Grunt, *and now this important book. I am so grateful.*

~ Bob Fischer

I dedicate this book not only to our nation's veterans but to my wife and my sons, who endured the good and bad times of a Marine Corps Vietnam veteran. I am forever grateful for your love and support.

~Grady Birdsong

Most significantly, this book is dedicated to all men and women who volunteer to serve and fight for our country. It is these unknown, unsung patriots, who never receive the credit they deserve, that should be honored by way of this book. It is because of these heroes, who fight in the trenches, in the air, and on the seas, that we citizens of the United States of America are able to enjoy life and carry on our daily lives in civilization's greatest experiment of the world's greatest government—and greatest economic and military power ever.

CONTENTS

FOREWORD

It was 2008 and we'd begun to see an increasing number of military members return from Iraq and Afghanistan with traumatic brain injury (TBI) and post-traumatic stress disorder (PTSD). The Airman who sat in my office appeared quiet, a bit sullen, and apprehensive. His wife was by his side to offer encouragement and support. He needed it.

The Airman's commander had recently told him he was on his way out of the Air Force, facing a medical evaluation board (MEB) and maybe disciplinary action. He had been injured in an improvised explosive device (IED) blast in Iraq, along with his buddy, and was having trouble doing his job as a truck driver. He had headaches, couldn't sleep, and was irritable and forgetful. He couldn't remember his truck routes or lock combinations. He was making mistakes at work and getting into trouble for it. The sleeping and headache pills he got from the clinic weren't helping.

He could date the onset of his problems back to the IED blast and tried to explain this to his docs, but they dismissed what he told them and diagnosed him with migraines and PTSD. The Airman also had a TBI. Although I had "official" responsibility for this Airman, I knew he would be best served by being treated off-base. I reached out to a friend of mine in the private sector, Dr. Albert "Eddie" Zant, who had seen an article in the local paper about this Airman. The Airman had been awarded a Purple Heart for his injuries. Dr. Zant said he would treat him without charge, so I referred him to Dr. Zant's private practice for hyperbaric oxygen treatment. As the 720th Special Tactics Group Surgeon, I didn't have any real authority over what happened to the Airman, but I knew the regulations, and I knew about hyperbaric oxygen. I called the Airman's commander and told him to put the MEB on hold, stop any discharge and disciplinary action, and give us the opportunity to try to get him well.

Dr. Zant treated the Airman and his buddy, the other Airman who had also been wounded, and they both got well. By the time the Airman completed the treatments and he sat across from me in my office once more, I saw that he had a renewed sense of hope and encouragement. When he left my office, he was smiling for the first time in a long while.

Both Airmen returned back to their Air Force careers healed. That was my first experience treating TBI with hyperbaric oxygen, and I became a believer.

The truck damaged by the IED in which the airmen were riding

The Miracle Workers of South Boulder Road is a story of hope, courage, persistence, and healing. There is no way anyone can put a price on a life regained, a marriage saved, or a career renewed. These are true miracles and each one is an uplifting story. If it were not for the faith and perseverance of Ryan Fullmer, Eddie Gomez, Pepe Ramirez, and Dr. Julie Stapleton, along with their wonderful team of workers and supporters, these miracles would not have happened or continue to be taking place.

As I write these words in 2016, however, there are still plenty of medical professionals who say "hyperbaric oxygen does not work" for TBI. I wonder what these naysayers would say in front of one of the many recipients of hyperbaric oxygen treatment who got a life back, became well, was able to return to school, get a degree, or become a loving and compassionate spouse and parent again? Instead of telling TBI and PTSD victims that "this is the way you will be for the rest of your life," the Boulder team dispenses hope and healing. What more can one ask of a medical professional?

I hope you are as encouraged and hopeful as I was after reading this book. May God continue to bless the Miracle Workers of South Boulder Road!

~ James K. Wright, MD, Col, USAF (Retired)

Dr. Wright is a 1974 graduate of the University of Chicago, completing his surgical internship and residency at the University of Chicago (1974 -77) and UCLA (1977-79). He completed fellowships as a Clinical and Research Fellow (Burns) at Harvard Medical School in the Shriners' Burn Institute, Boston, MA, in 1975 and Hyperbaric Medicine at Brooks Air Force Base, TX, 1999–2000. There, as an Air Force Colonel, he was the Chief of Hyperbaric Medicine Research at the Davis Hyperbaric Laboratory. In conjunction with his personal friend, Dr. Eddie Zant, MD, then in private practice, they successfully used hyperbaric oxygen treatment therapy to treat and heal fifteen airmen who suffered the signature wounds TBI and PTSD.

Commissioned as a captain in 1982, he served in a number of capacities in his career that include Air Force pilot, surgeon, educator, and researcher, and was one of the pioneers and investigators in a national, multi-site clinical trial using hyperbaric oxygen for TBI. He left the Air Force to enter private practice in Midwest City, OK, in 1989 and subsequently entered the Air Force Reserve. After serving as plastic surgeon for Tinker AFB, OK, he also attended the Aerospace Medicine Primary Course. Col. Wright then re-entered active duty service and was assigned as chief of flight medicine at Tinker AFB, OK. He has deployed in support of contingency operations to El Salvador, Turkey, and Saudi Arabia. He is also the author or co-author of eleven publications.

Note: The Rocky Mountain HBOT Association for Brain Injuries also congratulates and supports Dr. Eddie Zant and his own miracle HBOT program that is based on the Israeli TBI treatment protocol. He has also treated a number of combat veterans and provided 90% of their fund support. Most of them return to active duty, private employment or are continuing their education.

INTRODUCTION

Professor Morris Massey of the University of Colorado says that our lives are interrupted and inspired by significant emotional events that change us forever. Mine certainly was in April 2010, when several of us attended the Rocky Mountain Hyperbaric Institute's presentation of its hyperbaric oxygen treatment (HBOT) program in Boulder, Colorado. Expecting another snake-oil pitch, it actually gave me that emotional event I had been looking for most of my life. Back then, this fledgling HBOT clinic was expanding its HBOT therapy program—that had up to this point been treating and healing civilian stroke victims—to include military combat veterans who were still suffering from the signature wounds of war: traumatic brain injury (TBI) and post-traumatic stress disorder (PTSD). Grady Birdsong and I were both Vietnam War veterans and had experienced our own share of combat stress, so we could relate to the PTSD cases. We quickly became veteran advocates for this program.

Ryan Fullmer had set up his "institute" in a rear office of a friend's stereo radio manufacturing shop in the industrial area of Boulder. Eddie Gomez, Ryan's newfound partner, had successfully established a 501c3 nonprofit for treating veterans—the Rocky Mountain Hyperbaric Association for Brain Injuries. They had just treated their first combat veteran, and several more were scheduled when Grady and I first joined him in his "broom closet" clinic.

Initially, Ryan established his HBOT clinic to treat and heal civilian patients, and they still comprise eighty percent of the Rocky Mountain Hyperbaric Institute's business. Most of these patients are treated for strokes, brain injuries suffered in accidents, wound healing, and other Food and Drug Administration (FDA)-approved conditions.

In 2010, the Healing Our Heroes program (and its nonprofit fund) was established and began to attract a large number of local donors to treat veterans with TBI and PTSD. Ryan and Eddie used the early funds Grady and I raised through the American Legion, Military Order of the Purple Heart, the U.S. Navy

Cruiser Association, the VFW, and our own Cooper's Troopers (Denver area Marine Corps luncheon group) to begin treating veterans regularly.

In the following pages, the miracle workers, Ryan and Eddie, will tell us their own stories in their own words, including how they each first became certified hyperbaric technologists and later certified hyperbaric specialists. Ryan began treating his first civilian patients in the two oxygen chambers he struggled to buy and install in the Boulder "broom closet." Recognizing that the FDA controlled and authorized the HBOT protocol and specified exactly what therapies or conditions they could treat, Ryan and Eddie walked the legal, medical, and technical gauntlet that allowed them to treat and heal many combat veterans ... over 300 as of this writing.

Hyperbaric oxygen chambers are now approved for use by the FDA, but only for the following treatment conditions:

(1) Air or gas embolism
(2) Carbon monoxide poisoning
(3) Gas gangrene
(4) Traumatic ischemia-crush injury
(5) Decompression illness – U.S. Navy "bends"
(6) Arterial insufficiencies
(7) Severe anemia
(8) Intracranial abscess
(9) Necrotizing soft tissue infections
(10) Refractory osteomyelitis
(11) Delayed radiation injury
(12) Compromised flaps and grafts
(13) Acute thermal burn injury
(14) Problem wounds
(15) Retinal ischemic disease

Nowhere in this grocery list of treatment conditions are strokes and TBI therapy approved of or even condoned by the FDA, nor is PTSD treatment even acknowledged. Yet, over 300,000 combat veterans desperately need TBI therapy and more than 600,000 suffer from PTSD. (See Exhibits A and B.)

You may be wondering, since the FDA does not approve the use of hyperbaric oxygen chambers for TBI or PTSD, how does a hyperbaric oxygen program like Ryan's and Eddie's even function, much less survive? It all depends upon the way it is employed and by whom. Strokes, TBI, and PTSD are all considered off-label conditions by the FDA, just as the majority of drugs administered by the Veteran's Administration are also considered off-label. However, within the FDA provisions of this off-label situation any medical doctor can screen veterans and write an HBOT prescription for treatment, as Ryan and Eddie's clinic provides. Dr. Julie Stapleton, Rocky Mountain Hyperbaric Institute's medical director, does just that. It is legal and well accepted, but only a licensed physician can write it—and that is how Ryan and Eddie are able to treat and heal veterans.

Fortunately, Ryan and Eddie recognized the necessity to include PTSD cognitive and associated physical therapy in their total treatment program. Today, the Rocky Mountain Hyperbaric Institute has the only fully integrated TBI-PTSD therapy in America, thanks to the skill and dedication of Sergeant Major Pepe Ramirez, a Marine combat veteran himself. He not only has the respect and confidence of these veterans but has the counseling credentials to heal them. He is currently working on his Ph.D. in Clinical Psychology with concentration in Neuropsychology. At the Rocky Mountain Hyperbaric Institute, our veterans have their TBIs healed and their psychological and mental wellness restored.

The most definitive discussion of PTSD has been provided by our guest authors, Xavier Figueroa, Ph.D. and James K. Wright, M.D., Colonel USAF (Ret), whose article in *Brain Health & Healing Foundation* in November 2014 is entitled, "OK Doc ... What Do I Really Have? PTSD vs TBI?" (www.brainjury.org). They make the important distinction between these signature wounds and confirm what our resident PTSD counselor looks for throughout the forty-hour treatment period: recurring recollection of the traumatic event, distressing dreams, withdrawal and avoidance of anything associated with the trauma—like people, places, and activities. He also monitors a veteran's withdrawal, lack of interest, memory loss, as well as their grief, guilt, angry outbursts, lack of sleep, startled responses, and signs of hyper-vigilance.

Admittedly, as veterans ourselves, Grady and I have become strong advocates of the Rocky Mountain Hyperbaric Institute and its nonprofit Rocky Mountain Hyperbaric Association for Brain Injuries. However, it is not the HBOT therapy itself that has attracted veteran advocates like Grady and me and the very large donor group that continues to support this special program. It is as a result of witnessing the remarkable transformation these veterans undergo during their forty chamber hours at the clinic. We have had the opportunity to meet and be of support to "basket case" combat veterans who are first diagnosed with severe memory loss and the inability to sleep normally, and who are struggling with ongoing debilitating and excruciating migraine headaches. Sometimes, they are also dealing with a marriage that is either in crisis or completely broken or a family that cannot understand the veteran's withdrawal and suicidal tendencies. Virtually every combat veteran who suffers improvised explosive device (IED) blasts and concussions exhibits these symptoms upon his or her arrival for HBOT treatment. And with almost every TBI case, there is also the accompanying, interwoven PTSD that further complicates successful treatment. After their therapy, we have witnessed firsthand how their bodies, lives, and relationships are healed.

Since 2010, Grady and I have met and supported most of these veterans who received treatment. It has been an honor and a privilege to know each one of them, as well as many civilian patients who have been treated. Although every individual's progress is significant, there are a few cases that especially stand out in my mind and reflect how credible and successful HBOT therapy has become.

October 2010: Fidel G.

Staff Sergeant Dean Sanchez, the U.S. Marine Corps Wounded Warrior Representative for Colorado, called to ask if we could help Fidel, who he found living under a bridge in Colorado Springs. He was suffering both TBI and PTSD symptoms. But to make things worse, he was partially blind in one eye and could barely speak. He had also been medically discharged with no VA disability or benefits. He was so incapacitated that he was unable to contact a Veteran

Services Officer, who assists the disabled at the VA. Staff Sergeant Sanchez arranged this critical support for Fidel, who then received a sixty percent VA disability rating. This action by Sanchez ensured that Fidel could receive medical treatment and prescriptions at the VA Hospital in Denver.

Along with Sandy Flint, Chairman of the American Legion's Wounded Warrior Program, we were able to find an apartment and feed Fidel until his application for HBOT treatment was approved in November. For the duration of his forty-hour treatment, Fidel lived in the Silver Saddle Motel, which was near the Boulder "broom closet" clinic. When his girlfriend came to care for him, she discovered that Fidel had taken a very bad fall and had broken his ankle. It necessitated surgery that was performed at the VA Hospital. The surgeon repaired Fidel's ankle and bandaged the five-inch incision he'd been required to make.

When Fidel came to the clinic, it was obvious that the bandage had not been changed and the wound was badly infected. Eddie Gomez, who is a certified nursing assistant, re-bandaged it before they began Fidel's HBOT therapy. To the amazement of Eddie and Ryan, the messy wound healed completely after just five chamber hours. Fidel also had his eyesight and speech restored to normal. He returned to his main interest in life, artistic welding, with no signs of his former TBI or other physical problems.

May to July 2013. Von Troung (civilian)

One of the eighty-percent-civilian population who received HBOT therapy at the clinic, Mrs. Troung suffered a serious stroke that left her right side paralyzed. She was unable to move her limbs. Three months of regular stroke therapy failed to give her any improvement. At that time, she heard about the Rocky Mountain Hyperbaric Institute and learned that the clinic treats and heals a large number of civilians who have suffered strokes and are brain damaged as a result of automobile and other accidents.

This sixty-year-old lady was one of the most experienced acupuncturists in Denver, Colorado, and she trained many of its doctors and nurses in this modality. Alternative medicine was not new to her when she applied to the clinic for her own HBOT treatment. Forty chamber hours later and commended as one of the clinic's best patients, Von continued her physical therapy and returned to work in her own clinic a year later.

March to July 2015. Joe J.

Navy-Marine Corpsman Joe J. was one of the most severely wounded combat veterans the clinic has treated and healed thus far. He joined the U.S. Navy in 2004 and graduated from Hospital Corpsman "A" School and the Field Medical Service School. In 2006, he was assigned to the 1st Marine Division's 5th Regimental Combat Team 5 that deployed to Fallujah, Iraq. Surviving dozens of dangerous missions, where he saved numerous lives and limbs, Joe was hit by nine IED blasts during that tour and was evacuated with TBI and PTSD. Treated for his signature wounds until 2009, he was subsequently medically retired as unable to perform his required duties.

This former straight "A" high school student suffered some of the worst TBI symptoms and PTSD that Eddie and Ryan had ever experienced with their first 100 veterans. It took two separate forty-hour treatment periods to return this dynamic young man to his former self, physically and mentally. His cognitive testing results, analyzed by Dr. Xavier Figueroa, showed the greatest improvement of any other veteran previously treated and tested at the clinic. In our last discussion with Joe, as he packed for home, he told us that he had applied to medical school.

Like Mrs. Von Truong, Ryan has attracted a large number of civilian patients who are treated and healed not only for the FDA approved conditions but also those suffering from strokes and brain injuries. He provides the same HBOT treatment plan and protocol for these civilians as he does for his veteran clientele.

Several of the most successful treatments include a two-year-old girl who had a serious brain injury from an accident that resulted in her being so spastic with involuntary movements that her flailing arms had to be strapped to her body at all times. Her mother had to hold her in the chamber during treatment. About one year later, this same little girl was running up and down the halls of the clinic, playing happily with a dog.

A second civilian patient was a professor at a local university who had been diagnosed with deadly ulcerative colitis that infested twelve feet of her colon. Doctors wanted to remove the diseased section by surgery. The operation would have left her with major digestive problems as well as other issues. Suffice it to say, Ryan's 100-hour HBOT program, with a few follow-up "maintenance" visits over the period of one year, totally healed her malady. "It was a miracle!" she proudly stated.

There were also a large number of civilian stroke victims, including our Von Truong, who came on crutches, were pushed into the clinic in wheelchairs, and even rolled in on gurney carts. They left standing, walking, and talking again. More miracles!

More than 250 veterans later and well-established in the Professional Medical Center, in Louisville, Colorado, Ryan and Eddie still inspire and sustain our *significant emotional event*. Ryan Fullmer and Eddie Gomez are the Miracle Workers on South Boulder Road who treat and heal combat veterans suffering from TBI and PTSD. And they do it using use one of the most common elements found in nature: oxygen.

Hyperbaric oxygen therapy is the *miracle* process that Ryan and Eddie both employed to treat and heal their own severe debilitating strokes before providing this same therapy to heal the hundreds of veterans who come to them with the signature wounds of war. Just as commendable is their Healing Our Heroes nonprofit fund that provides free treatment, counseling, housing, and food for these wounded warriors who are then able to restore their upended lives and the family support they lost when their physical and mental faculties were almost destroyed. A miracle does take place. Just ask their spouses, kids, and families at home.

~ Colonel Bob Fischer

"We have the most unique hyperbaric team in the country, which provides the integration of both traumatic brain injuries and post-traumatic stress expertise that treats an extensive local and national combat veteran population."

~ Eddie Gomez, President of the nonprofit
Rocky Mountain Hyperbaric Association for Brain Injuries

CHAPTER 1

Brain Injuries –
Common Occurrences

Traumatic brain injuries unfortunately are common occurrences and are ranked as one of the most prevalent injuries in our society today. TBI, as it is termed, is a major public health concern, especially in younger males and people in their elder years. Some of the most common causes of TBI for civilians include falls, motor vehicle crashes, being struck by objects or hitting against an object, and assaults. In 2010, 2.5 million TBI-related emergency room visits, hospitalizations, or deaths were recorded (Source: Centers for Disease Control and Prevention).

The Signature Wounds of the Current Wars

TBI-PTSD Signature Wound logo design
~ courtesy of Rick Baum, USMC Veteran, Minister & Veteran Advocate for the Rocky Mountain Hyperbaric HBOT clinic

TBIs have also become one of the signature wounds of our combat veterans in military deployment in the Middle East. Of the approximate 2.5 million U.S. troops who deployed to Afghanistan or Iraq, it is estimated by study teams that a significant percentile suffers from some degree of concussion and post-traumatic stress. TBIs, PTSD, and accompanying depression generally follow deployment, and almost thirty percent of combat veterans experience both TBI and PTSD, the signature wounds of our present-day wars—and what yesterday's combat veterans knew as "shell-shock" and "battle-fatigue." Unfortunately, most of these wounds remain untreated and misunderstood.

IEDs, one of the more deadly weapons devised, cause scores of mild to severe traumas for returning combat veterans. Those veterans who experience these wounds are all too familiar with IEDs and other concussive blasts. The author's son, Marine Sergeant Shane Birdsong, an infantryman, recorded two such blasts while on his first tours in Iraq.

Sergeant Shane Birdsong in Afghanistan
~ courtesy of Shane Birdsong, USMC

The photo below (Figure 1) shows a controlled blast by Marine combat engineers to destroy enemy ordnance found in an area while patrolling. This particular photo shows the magnitude of explosives in an IED cache that are typically discovered.

The next photo (Figure 2) depicts a less fortunate mobile patrol in Anbar Province, Iraq. This squad, while on patrol, took the full impact of a buried command-detonated IED that destroyed the vehicle and wounded all occupants—the driver, the squad leader, the A-gunner, and eight Marine infantrymen in the rear of the truck. All were medically evacuated with moderate to severe physical wounds and TBIs. All the Marines survived. The IED consisted of an explosive wired to a 155 artillery round and propane tank. The propane tanks used in these IEDs are like a normal American BBQ grill propane tank, which can produce horrendous concussive force if ignited by explosives.

Figure 1. Concussive blast in Iraq
~ courtesy of Sergeant Shane Birdsong, USMC

Figure 2. IED blast damage in Anbar Province, Iraq
~ courtesy of Sergeant Shane Birdsong, USMC

Brain damage from such blasts results in both physical and psychological injuries that regular, pill-driven medicine attempts to treat and heal. Yet, rather than prescribing drugs, we believe there are better ways to address these injuries, which is what is being done at the Rocky Mountain Hyperbaric Institute. Before we continue further to present the benefits and success of HBOT, however, it is important to first consider the brain and its complexity. Only then can we examine just how oxygen treats and heals TBI and also facilitates the treatment of PTSD.

The Human Brain: Injury and Recovery

The most fragile, complex, and vulnerable organ in the body is the human brain. It controls our thoughts, emotions, behavior, movements, and sensations. It processes what we think, feel, and create with its billions of neurons and their dendrite forest. They work together to receive and exchange data and then act as the interpreter of our senses to control and direct all our movements. Even our sleep and breathing depend on this healthy condition. Each of the brain's five lobes has specific functions that we depend on to live, function, and experience life.

The brain's gelatinous mass is encased in a hard skull for protection (Figure 3). It wards off mild concussions and blows to the head that injure us when we fall or "have our bell rung" in a boxing ring. In his description of the human brain and how it functions, hyperbaric treatment pioneer Dr. Paul Harch employs a brain-sized stalk of broccoli (Figure 4), as he shows its many thousands of tendrils and then describes how a brain's dendrite forest reacts to a hard blow or a serious IED blast (see Figures 1 & 2).

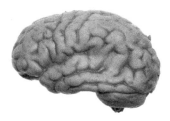

Figure 3.
The human brain

Figure 4.
Broccoli stalk

If we carefully examine broccoli we see these many tiny tendrils (Figure 4) that are similar to the dendrite forests of the brain (Figure 5). The brain has billions of neuron cells that anchor their branched projections, the dendrites. They, in turn, propagate—sending and receiving electro-chemical stimuli—and act as antennae or receptors of signals from other nerve cells.

The human body and all its physical and emotional functions depend on the healthy performance of these critical brain elements. When they are damaged from a blow or concussive brain injury they either die or shut down. Recent brain studies have confirmed that the brain has an amazing capacity to recover and the brain's own plasticity generates brand new dendrites that replace the dead ones or revitalize the dormant ones. Spectrographic scans taken of a veteran's concussion-damaged brain show areas of injury to the lobes and their dendrite forests. As we confirm with our HBOT treatment, concentrated oxygen not only treats and heals a damaged brain but stimulates and activates the brain's own ability to heal itself. Recent Israeli neuroplasticity studies of veterans with TBI confirm this process.

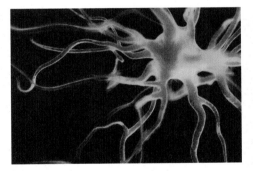

Figure 5.
Healthy Dendrites

Figure 6.
Damaged Dentrites

The Veteran's Brain Suffers a Concussion

The brain is composed of eighty billion dendrites that are anchored in their neuron bodies. They are the primary receptors for our brain's information and together they control and coordinate our body's many functions. In effect, they are the main elements of the brain's computer.

The human brain is designed to ward off minor blows, such as from a fall or as a result of being shaken in an accident. The brain mass actually bounces off the inside skull wall and in most cases is saved from any injury. A serious blast or concussion, however, sends shockwaves through the brain mass with such intensity that many dendrites and neurons either shut down or die as a result of the severe shock. Oftentimes, these dendrites survive serious damage and go dormant until the proper type of medical treatment revives and restores their lost functions. (Figure 6).

For several centuries, it has been proven that oxygen heals wounds and destroys parasites. Oxygen has been employed for more than 100 years by the U.S. Navy to treat and heal nitrogen narcosis, or the "bends," as it is better known. Major hospitals today employ hyperbaric chambers for a large number of FDA-approved maladies, as shown in the following list: air or gas embolism, carbon monoxide poisoning, gas gangrene, crush injury, decompression sickness, arterial insufficiencies (category of diabetic foot ulcers), severe anemia, intracranial abscess, necrotizing soft tissue infections, osteomyelitis, delayed radiation injury, comprised skin graphs and flaps, thermal burn injury, and idiopathic sudden sensorineural hearing loss.

An important Israeli study, documented and published in 2013 by a distinguished team of researchers drawn from Tel-Aviv University, The Assaf Harofeh Medical Center, and the Institute of Hyperbaric Medicine, Zerifin, Israel, tested and documented the brain's capacity to heal itself in at least two cases.[1, 2] Called neuroplasticity, it is the brain's unique ability to overcome and eliminate injuries and tissue damage with the right stimulus, such as the brain's exposure to hyperbaric oxygen, which actually heals and restores life to damaged cells.

Blood flow is critical to human life and bodily functions. The brain is no exception, and its own complexity requires even greater attention than other organs and wound areas. Interruption of blood flow to parts of the brain can result in strokes, seizures, and other impairment, which is why the timely HBOT treatment of veterans who suffer from TBI and PTSD is extremely critical, especially when there is proof that oxygen-based therapy really does heal and restore the combat veteran's body and mind.

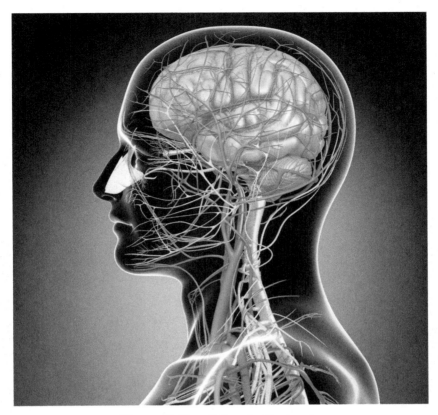

Figure 7. Blood flow in the brain

How New Blood Cells and Pathways Are Created

An adequate supply of oxygen and nutrients is critical for all human cells and tissues. Nurturing the proliferation of the below processes with pressurized oxygen has been found to accelerate healing of traumatized areas in the brain.

Angiogenesis: The development of new blood vessels from pre-existing vessels. It occurs in the healthy body for healing wounds and restoring blood flow in damaged tissues after injuries.

Neovascularization: The formation of functional microvascular networks (blood paths) in red blood cell perfusion. It differs from angiogenesis, which is characterized by profusion and outgrowth of capillary buds and sprouts from preexisting blood vessels. Hyperbaric oxygen stimulates both.

The Injured Veteran's Optimal Course of Treatment

Each veteran who is treated at the Rocky Mountain Hyperbaric Institute receives a concentrated oxygen brain bath for forty hours, one hour at a time. The following schematic shows what happens inside the damaged brain when oxygen is introduced at 1.5 atmospheres (the equivalent of 17 feet below sea level at a 5,000-foot altitude). The oxygen healing process includes both angiogenesis and neovascularization, as defined above. The damaged dendrites and neuro pathways are either bypassed, if the cell or dendrite is dead, or can be revitalized if in a dormant state (Figure 8).

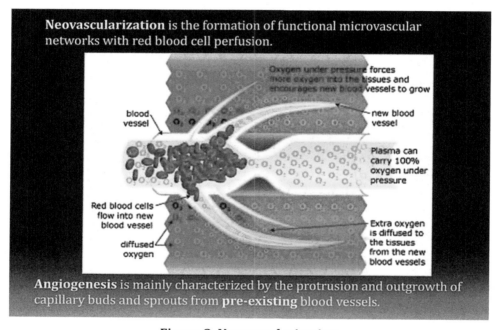

Figure 8. Neovascularization

While there may be few exceptions, virtually every combat veteran who suffers from TBI also has PTSD. The *American Journal of Psychiatry* and the Mayo Clinic have a similar definition: "Post-traumatic stress disorder (PTSD) is an anxiety disorder that may develop after exposure to a terrifying event or ordeal in which severe physical harm occurred or was threatened. Traumatic events that trigger PTSD include violent assaults, natural or unnatural disasters, accidents or military combat."

They further define the signs and symptoms of PTSD and group them as follows:

1. **Re-experiencing symptoms:**

   - Flashbacks – reliving the trauma over and over, including the experience of physical symptoms like a racing heart or sweating. Words, objects, or situations can trigger these symptoms.

   - Bad dreams and frightening thoughts that may cause problems in everyday life.

2. **Avoidance symptoms:**

   - Staying away from places, events, or objects that trigger experience reminders.

   - Feeling emotionally numb, with strong guilt, depression, or worry predominant.

   - Little interest in activities that were enjoyable in the past—before the trauma.

   - Having trouble remembering or recalling the dangerous or terrifying event.

3. **Hyperarousal symptoms:**

   - Easily startled, to the extent that a person may change their personal routine.

   - Feeling tense or "on edge," which can lead to more stress and angry outbursts.

   - Inability to function normally to do basic daily tasks, usually with little sleep.

When the injured veteran arrives for treatment, it is common to find that they suffer from many of the symptoms listed above. Their disrupted sleep, serious memory loss, mood swings, and withdrawal-depression require the immediate attention of the clinic staff. Many do not communicate at all or lack the ability to speak clearly or even pay attention to their first instructions.

They must be counseled and gently led through each TBI and PTSD treatment step until each of them becomes familiar with and trusts the clinic's medical director, the HBOT technicians, the PTSD counselor, and the chamber operators. At first, it is a foreign environment and can be quite intimidating.

★ ★ ★ ★ ★

Today, more than 250 healed veterans can now attest to the professional and successful TBI and PTSD therapy they experienced in the only integrated TBI and PTSD treatment program in America. When each veteran started the program, it was obvious that they suffered some degree of brain damage from their combat concussions or other injuries. In too many cases, they had suffered their signature wounds for far too long and their families and friends had suffered along with them. Their brain injuries went virtually untreated and unhealed until they sought help at the Rocky Mountain Hyperbaric Institute. When they arrived at the clinic, a large segment of them were actual "basket cases." Most were doped up on useless off-label medicine, aka "black drugs," the majority of which had not yet been approved or tested by the FDA. Some veterans arrived with many different prescriptions, as if more of these pills would somehow help them feel better.

The need for additional health services, especially for neurotrauma, has skyrocketed in recent years. And it likely will increase even more significantly as time goes forward. Invisible wounds have become the signature wounds of this generation and these combat veterans. This brings us to our story ... the miracle that is happening on South Boulder Road in Louisville, Colorado. Ryan Fullmer, the program director, Rocky Mountain Hyperbaric Institute, has partnered with Eddie Gomez, President, Patient and Nonprofit Services, Rocky Mountain Hyperbaric Association for Brain Injuries, to treat and heal traumatic brain injuries. They also integrate PTSD counseling within the same TBI treatment period, usually over a period of forty days. This is one of the most imaginative and successful programs of its kind in the United States and combines the treatment of the two maladies under one comprehensive treatment process.

Before we move into some of the amazing stories of recovery that have occurred at the Rocky Mountain Hyperbaric Institute, we'd like to first introduce you to this miracle team.

CHAPTER 2

Ryan Fullmer –
In His Own Words

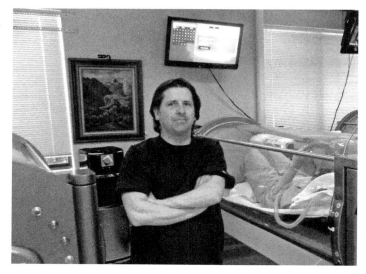

Ryan Fullmer, Director of the Rocky Mountain Hyperbaric Institute
~ courtesy of Grady Birdsong, Veteran Advocate

Upon entering the Rocky Mountain Hyperbaric Institute, it is easy to misjudge who is in charge, with the hustle and bustle of patients, technicians, and family members milling around. Yet, that first impression of this clinic also reveals the efficient and orderly management of two types of clientele, the civilian patients and the military veterans who seek treatment for their post-service injuries.

The first impression one may have of Ryan Fullmer is that he is another patient in the clinic, since his quiet, unassuming, low-key manner and attire mask his astute management style; that is, until he spots a problem or senses he must intervene to keep the daily activity on schedule. Only then does it become apparent that he is the director of this unique business.

His position also becomes evident when he briefs a new patient or a confused veteran for the first time. He quietly explains the program and shows them what they can expect when they enter the hyperbaric chamber in his sparkling clean chamber room.

At times, you may also find him presenting his introductory video to a visitor or a family member interested in what the process offers. But he is at his best when called upon to describe the hyperbaric process and what happens during healing of a civilian patient or wounded warrior who suffers from TBI and PTSD. Ryan Fullmer is a unique individual, and so is his Rocky Mountain Hyperbaric Institute. His clinic treats and heals the whole spectrum of these signature wounds, both physically and mentally.

Ryan is a master of his workplace. His knowledge of HBOT is almost dumbfounding, since there is virtually no technical detail and no aspect of it that he cannot explain in clear and concise detail. The recipient leaves with an education, insight, and confidence that such a dedicated individual is handling their care or the care of a loved one or friend.

Likewise, when Ryan presents to veterans' groups, or perhaps a Rotary club, he answers technical hyperbaric questions with ease, as if he had written the book on this complex subject. It is then that others become enthused with his dedication and knowledge of hyperbaric therapy and its proven healing results.

When did this modest, unassuming individual become interested in such a different treatment program as hyperbaric therapy? Certainly, he did not learn about it in America's medical community, where hyperbaric oxygen treatment ranks low—along with acupuncture, chiropractic, and other alternative medicine therapies.

How did a seriously disabled young man survive his own severe stroke and then take on a huge challenge that would result in helping so many other lives and their futures? Ryan's personal story is a remarkable one.

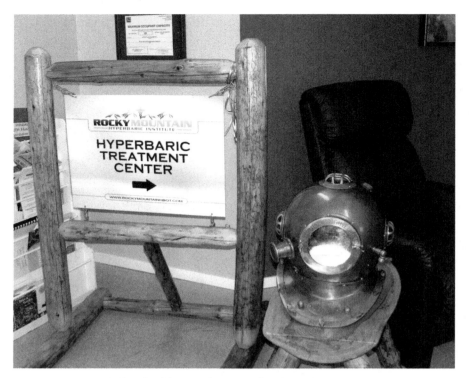

Rocky Mountain Hyperbaric entrance
~ courtesy of Grady Birdsong, Veteran Advocate

Ryan's Journey

Ryan, born in April 1975, is the third son of four boys to Jon and Pat Fullmer. At the time, the Fullmers lived in Salt Lake City, Utah. His mother's family originated from this area. Ryan's mother was active in the church and cherished being close to her large family. Ryan remembers that Christmastime "was my best memory of living in Salt Lake, with tons of family and good times."

Ryan's father, Jon, was beckoned by his father to move to Gunnison, located in the mountains of central Colorado, to help with the family hardware store. Gunnison is well-known for its long winters, snow, and cold weather. "Much to my mother's chagrin, we packed up and moved to Gunnison, only to arrive during a blizzard with a lot of snow on the ground. We did find eventual relief from that boredom and began sledding, skiing, snowmobiling, and doing all those really fun snow activities."

Ryan, brother Eric and Dad, Jon Sr., skiing
~ courtesy of Patricia Fullmer, Ryan's mother

Ryan had a typical childhood in Gunnison, where he participated in Cub Scouts, baseball, and piano lessons. Education was equally important, because Ryan's mother was an English teacher and his father was an accountant. Poor grades were not tolerated. Both parents were very involved in their sons' educations, which included attending church regularly. When the boys were not in school or working part-time at the hardware store, they were assigned various duties, like taking out the trash, chopping wood, or shoveling snow. "I went to school, then worked at the store in the evenings and weekends. Whenever I would beg for an allowance my father would say, 'You don't need an allowance, you have a job.'"

**Fullmer family of Gunnison, Colorado, 1988 – Upper LR:
Robert, Jon Sr, Patricia, John Paul Lower LR: Ryan & Eric**
~ courtesy of Patricia Fullmer, Ryan's mother

"I was a typical young kid, in good health, with a normal routine growing up … nothing out of the ordinary. Right after my eighth birthday, I attended a friend's birthday party. He'd invited some of our school buddies to help him celebrate. It is hard to remember everything that transpired at the party, because this happened thirty years ago. I do remember the mountains of birthday cake, hotdogs, and presents. After the party, some of us continued to hang out and played Donkey Kong on the Atari. Suddenly, I came down with a wickedly bad headache—one like I had never experienced before. I didn't want to quit, so I kept playing the game. The headache then started to become more intense. *Was it something I ate?* I wondered. 'Chris, come help me get up,' I gasped out to my friend. The headache had become almost unbearable. Try as I might, I could not stand up. *Why won't my arm work?* I thought."

Ryan Fullmer, age 7
~ courtesy of Patricia Fullmer, Ryan's mother

Ryan Fullmer's recollection over thirty years ago becomes real again as he recalls his first "brain attack." The next few days were a blur, filled with the whirling activity of multiple doctor visits and excessive sleeping. Physically, Ryan struggled with the simplest movements as he tried to stand up, fell over, and tried to stand up again. "I was not able to walk. My left arm was totally unusable."

Healthcare in Gunnison, Colorado, in 1983, was adequate but had not seen specialization care yet. Although the ER doctor in Gunnison didn't know specifically what was causing Ryan's condition, he sent him home with instructions for his mother to "Give him aspirin and he'll be fine." Ryan's mother was not satisfied with the doctor's inability to figure out what was wrong, so after a few days she took Ryan back to the emergency room. During this second visit, they diagnosed Ryan with brain cancer. "Not the kind of news a parent wants to hear about their young child," Ryan reminisces.

The hospital policy required that patients like Ryan be airlifted to a more specialized facility. That facility was in Grand Junction, Colorado, and a raging blizzard was in progress. Ryan was hastily taken by ambulance to Montrose, Colorado, just south of Grand Junction, and then airlifted to Saint Mary's in Grand Junction. During this time, Ryan lapsed in and out of consciousness. After a series of CAT scans, it was ultimately determined that little Ryan Fullmer had suffered a massive ischemic right hemisphere stroke. An ischemic stroke, or "brain attack," is caused by a blood clot that interrupts the blood flow to the brain. In Ryan's case, he had lost the whole left side of his body.

The Rocky Mountain HBOT crew and their patients
~ courtesy of Ryan Fullmer

The next few years were filled with a series of tests, in-patient stays, physical therapy, and medical evaluation. "I became a guinea pig in a lab. Often I would hear, 'An eight-year-old shouldn't have a stroke. That is an old-person injury.'" Ryan was destined for many challenges during his childhood. Slowly, Ryan began to walk again but had not regained the full use of his left arm by the time he was ten years old. "It was a tough time for me."

Ryan continued going to physical therapy. He recounts those days, "My parents kept pushing me to do physical therapy. Of course, I still went to school every day. You see, they never treated me like I was defective or different from the other kids. When I graduated from high school, I decided to attend Dixie State University in St. George, Utah. During that time, I received a tendon transfer procedure on my left arm and, after recovering from the operation, for the first time in twelve years my arm was somewhat useable." It wasn't until later in life that events began to unfold in even more hopeful ways.

Breakthrough

"New Hope for Stroke Patients" was the newspaper headline. A friend had found the article in the *Denver Post* and alerted Ryan. Ryan immediately called about the seminar that was being held in Denver; he was the third person to answer the call for stroke patients. His mother went with him to Denver to attend a seminar on hyperbaric oxygen therapy. Ryan found the seminar very exciting, and it filled him and his family with hope. "I was the second person to sign up for treatment. The positives were that it was safe, effective, and had no side effects, but the downer was that those forty treatments would cost $20,000. To top that, I did all of the treatments in forty consecutive days."

Ryan, with new enthusiasm, found a place to stay and settled in for the treatment process. He showed up with headaches, depression, limited use of the left side of his body, and a brain that was foggy. The staff thought that his stroke, which had happened fourteen years prior, would not be helped. However, they were willing to give his case a try. Ryan had nothing to lose, and within the first few treatments, his headaches started to fade, the stiffness in his left side relaxed, and he immediately experienced further improvement when his foggy brain activity disappeared.

This was a revelation to Ryan. "My left ankle had not moved in fourteen years, and within the first two sessions my ankle started working little by little." He began to walk better and more steadily and felt that the physical therapy was worthwhile. After he was there for a while and began to feel good about the improvement he was experiencing, he recalls, "I started studying everything I could get my hands on in the clinic that pertained to hyperbaric oxygen. I spent all day at the clinic helping out where I could. I would talk to patients, do laundry, clean the area, and read everything. I was a sponge, and for the first time in a long time the brain fog was gone and I could retain what I was reading."

The clinic owners needed help, since this was a startup operation, so they hired two women from Florida. One of the women was a certified hyperbaric technologist with a lot of experience and the other was a registered nurse. They also happened to be mother and daughter. Ryan, as a patient, naturally gravitated toward them because they were eager to help and willing to teach him more about this new treatment. Most importantly of all, Ryan noticed that all of the patients going through the program were getting better. Then a true miracle happened for Ryan.

The Beginning of a Miracle

One of Ryan's most vivid memories at this clinic began when he met a patient who had been shaken violently as a baby and suffered a severe traumatic brain injury. That patient was in a wheelchair and was Ryan's age—twenty-two at the time. He could not walk, use his arms, stand on his feet, or feed himself with a knife or fork. He spoke in broken tone patterns that only his mother could understand. "It was tragic and heartbreaking for me to witness his condition and to hear that the staff did not think this treatment would help him."

To everyone's amazement, this fellow started to improve. He started swallowing better, making eye contact, and talking a little more clearly. He even began using his arms. Ryan reminisces, "Keep in mind that this was a person who'd been written off by doctors and neurologists. Nobody knew what to do with this guy. In twenty-two years he had never improved. I felt for him because, like me, nothing had worked for him up until that point. In time, he could feed himself and started to speak better. He began smiling. He even recognized me and spoke my name. I was there the day he took his first step … most of us had to compose ourselves. I was forever changed!" All of the patients who had signed up for this program were showing remarkable progress.

Ryan finished his forty treatments and was feeling great. He headed home to Gunnison to go back to work. However, his homecoming became bittersweet and was much different than before he left. Even though he loved working at the hardware store and being around his family, his head was still back at the clinic in Denver. He really wanted to do more treatments and work with the people he had met at the clinic. Counting nuts and bolts and stocking shelves seemed mundane after the discovery of, and his experience with, hyperbaric therapy. On blind faith, Ryan unceremoniously turned in his resignation to his father and returned to Denver.

When he showed up at the clinic, Ryan wasn't entirely disappointed that they weren't hiring. He agreed to exchange work for treatments. Cleaning, doing laundry, and running errands kept him in the game. He did not mind doing the menial chores because he felt that he had found his place in life. He was doing what he loved and was feeling better and better as time passed. Ryan remembers, "One day a new patient was starting. He was an older brain-injured, retired Army Vietnam combat veteran. He was in bad shape and was really scared about going in the "tube" [chamber]. After several failed attempts by the staff to get him in, I volunteered to work with him. Because I had spent so much time inside the chamber, I basically knew everything about it and how to run it. The staff and his family were amazed at how I put him at ease, explained what was going on and why, and then talked him through the whole treatment process."

Soon after, Ryan was hired as a chamber tender and spent his days running six chambers, sometimes by himself. At that time, the clinic was a very busy center and doing approximately thirty to thirty-five treatments a day. "No doubt I was tired, but when you are doing something that you love, it doesn't seem like work."

A Game Changer

In a short while, Ryan enrolled at the College of Oceaneering to take their course in diving and hyperbaric medicine. This would require a year of study and a supervised internship of 500 hours. After completing all of these requirements, Ryan sat for the certified hyperbaric technologist (CHT) exam in Texas, passing it in October 2002.

After completing the CHT certification, the clinic where Ryan had been working closed its doors for financial reasons. With no other clinics hiring in the area, Ryan went back home to Gunnison and resumed his hardware store duties. Over the next few years, he began dabbling with several clinics in the Santa Fe, New Mexico area and in Salt Lake City. As Ryan recounts these experiences, "I really didn't care for the way these clinics were run, especially when it came to safety. Safety had been ingrained in me, and I would not compromise my safety practices, so it became difficult to find the right employment in this field that I loved so much."

Ryan continued to find ways to work in the field, even trying a short stint with soft side chambers, but this type of equipment never seemed to catch on. Ryan finally received certification as a safety director in hyperbaric therapy in 2004.

Tragedy Strikes

In 2007, two young boys were playing together behind their house in Gunnison, Colorado. Slipping through a hole in the fence, they both inadvertently fell into a ditch filled with swift-moving runoff water that was raging downstream. Both boys were found minutes later downstream—one was dead and the other barely alive. The boy who lived came out of the ditch alive, but with limited brain function. Ryan and the community were devastated. As Ryan tells the story, "The community was turned upside down over this tragedy. How could this beautiful child with God-fearing parents end up in such a compromised state of existence? Naturally, they sought my advice. I sent them to Dr. Stoller's clinic in Santa Fe, New Mexico, where they generously gave this young man a series of hyperbaric treatments. But such a severe brain injury can't be healed with forty or eighty treatments. He needed hundreds of treatments ... like me."

Ryan took it upon himself to try and find a way to help the boy. At the time it seemed that the only answer was to form a nonprofit and use the donations to procure a chamber. Then, as a CHT, he could give the young man the treatment he needed. In making contacts as he looked for a chamber and the means to start a 501c3 nonprofit, Ryan realized it would take years and much more money than he personally could raise to start his own clinic. Although he came to the realization that he could not start a nonprofit at the time, in doing his due diligence, Ryan became familiar with some of the big names in the hyperbaric community.

A Short Stint with Hyperbaric Therapy

One day, Ryan's phone began ringing off the hook. A wealthy Boulder, Colorado, home audio manufacturer named Charlie Hansen had been hit by a motorcycle while he was on a bicycle ride. As a result, his spine became crushed, and he was expected to be a paraplegic. He certainly did not want to be disabled. Charlie began doing research on spinal injuries and discovered that hyperbaric oxygen could be advantageous in this acute stage. He proceeded to buy a chamber from Hyperbaric Clearing House that was being shipped to him directly and knew he would need a certified person to show him how it worked. In asking questions about operating the chamber, Hyperbaric Clearing House gave him Ryan's phone number.

Charlie hired Ryan to come to Boulder for a week and train him and his people on the operation of the chamber. The relationship was certainly a good one, and as Ryan remembers, "I treated Charlie for a week, wrote their policy and procedures, trained one of his manufacturing workers, and went home. Every day they called me with one concern or another. Basically they were concerned about the oxygen being a fire hazard."

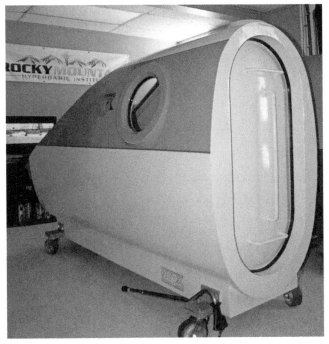

Second chamber in Charlie Hansen's manufacturing facility
~ courtesy Grady Birdsong, Veteran Advocate

The chamber became a bit temperamental. After weeks of constant phone calls, Charlie and his people realized they were in over their heads with the chamber and its operation. Charlie made an offer for Ryan to move to Boulder and treat Charlie once a day in the chamber. In exchange, they would let him use the chamber the rest of the time for his own patients. Ryan reflects, "Wow, what an opportunity and dream come true. *But, hell no*, I thought, *I can't stand to live in big cities! Where will I live? What about my beloved dog, Puppy?* I thought of a thousand excuses, including, *How can I quit my father again?* I was sure he was getting tired of the whole hyperbaric scene by then."

An Offer Not To Be Refused

Ryan resisted for a short time in the name of personal inconvenience. Every one of his protests was met with, "We are not going to take no for an answer. Keep throwing up excuses!" Ryan then fired back with a serious concern. This had to be addressed if a hyperbaric clinic could become operational. "I need a medical director who can write the prescription for hyperbaric treatment," to which Charlie replied, "Oh, her name is Dr. Julie Stapleton and she can't tell us no either!"

Like it or not, Ryan would have to leave his home, friends, and his family, and turn his small-town convictions into creating another life. Everything was in order for him—a chamber, installed and operational, real estate space, a medical director, and a lot of support.

This was one of the toughest decisions he had to face up to this point in his life. Ryan remembers, "Even though I swore that I would never move to the city to run a clinic, and against my family's wishes, I made the decision to do it. I would be starting all over again. I left so much in Gunnison ... you know ... not all risk is financial."

Dr. Julie Stapleton reluctantly agreed to refer three of her brain-injured patients to Ryan, and maintained that they had to show improvement after their hyperbaric treatment for her to stay involved. Charlie had set up this entire arrangement with Dr. Julie. That was in August 2007. It was agreed that if her patients progressed and showed evidence that they were healing, she would stay on with the program. Ryan now states, "Did it work? Did Dr. Julie stay on? She is most definitely still here and a pioneer in this hyperbaric treatment."

CHAPTER 3

Eddie Gomez –
In His Own Words

Eddie Gomez, President of the nonprofit
Rocky Mountain Hyperbaric Association for Brain Injuries
~ courtesy of Grady Birdsong, Veteran Advocate

Eddie Gomez is a fourth generation Coloradan. He proudly proclaims this and the fact that he was born at Memorial Hospital in Boulder, Colorado, which would play a pivotal role later in his life. According to his mother, Eddie was supposed to have been born on Thanksgiving. However, Eddie came into the world a week after Thanksgiving and tells people, "I guess I decided to let my family enjoy the holiday ... and I have always wondered if this might be the reason why I hate to be late for anything."

Eddie grew up third in the Gomez family of four children. He states that his childhood wasn't anything out of the ordinary. "We weren't a crime-fighting, superhero family, or any of those situations you see on TV, or any of the great things we dream up when we're children." He fondly remembers how his family always made trips to the majestic Colorado Mountains each summer. "The trip to Mesa Verde National Park and the Great Sand Dunes National Monument left a lasting impression. The trips to those places reinforced how beautiful my home state is and helped me make up my mind to stay here after college." The Gomez family, Eddie recalls, went on many outings and frequently took shorter vacations close to home at a local lake where they had a membership.

Idyllic Childhood

Eddie admits that until recently he never really appreciated that he grew up in the same house where his father had been born and raised. "Now, as this area grows, I have learned to appreciate how simple our lives were while growing up, knowing all the neighbors and graduating with many of the same friends that I started with in kindergarten."

Eddie and many of his childhood friends practiced and excelled in Little League football and baseball during fall and spring afternoons. They celebrated many championships together, gathering trophies as testament to those achievements. Eddie continued with sports activities throughout his elementary and middle school years and into his freshman year of high school. "It was my academic prowess rather than my athletic ability that would get me into college on a scholarship."

Eddie (right) at age 2 with cousin, Dax
~ courtesy of Eddie Gomez's mother, Cecilia

Eddie Gomez at age 8 played in Louisville, CO
~ courtesy of Eddie Gomez's mother, Cecilia

Eddie had played baseball and football since he was eight years old. He always held up his end of team responsibilities and was well liked by his close friends. He admits that he tried a little harder than most to be liked. "I always wanted to be part of the in-crowd, but the combination of being just okay in sports and being superb in academics made it a little hard to become noticed in the coveted social circles of the school. I was picked on by the older kids for being smart ... and being a little overweight marked me for occasional harassment."

Entering middle school, things did not change, academically speaking. It was here that Eddie began to understand his teachers' dedication to educating his young mind. This is when he began to fortify his mind toward a future teaching career. He recalls that most of his classes were extremely interesting, with the exceptions of art, shop, and home economics. It was at this point in his life that he started helping others in school. "I began to help my classmates grasp concepts and formulas ... the class courses helped me to stay challenged and the availability of honors courses made me even more determined. I definitely began to find my way, or so I thought."

Eddie (center) at age 10 with Cub Scout friends
~ courtesy of Eddie Gomez

Academics Come Easy

High school came quickly for Eddie, and he welcomed that transition with gusto. He immersed himself completely in his academic studies and left the sports behind. Recalling those fond memories, Eddie recaptured one of those moments, "Some of my former teammates asked me why I had quit playing. They wanted me to keep playing with them. Had I had that kind of encouragement while on those teams, I might have never stopped. But I knew in my heart that academics and a solid resume of associated activities were going to pave the way to college."

School was easy for Eddie. He was able to easily devour new material during school hours and got his homework finished before he left school. His love of most subjects kept him focused and involved in a lot of extracurricular activities. Having served with honors on the yearbook staff in middle school, those yearbook sponsors encouraged Eddie to write an introduction letter to the high school sponsors outlining his experience and expressing his interest in joining the staff as a freshman. "Much to my amazement, I was accepted on the staff … the first time a freshman was ever allowed to be a member of the high school yearbook team."

Eddie served as class treasurer his freshman and sophomore years. In his junior year, he ran for and was elected student body treasurer. In his senior year, he returned to the class senate and was elected to be the senior class vice-president. As a member of Key Club, a service organization sponsored by Kiwanis Club International, Eddie became a district lieutenant governor for the State of Colorado and helped develop high school Key Clubs in the district. Later inducted into National Honor Society, he served as president during his senior year.

Additionally, Eddie served as Spanish Club president, Student Accountability Committee president, and participated in numerous school plays, musicals, and show choir. As he tells about his extremely busy past, he gushes, "Whew! I guess you could say I came out of my shell!"

Eddie & Susan Stokes, Centaurus HS, Louisville, Colorado, show choir
~ courtesy of Eddie Gomez

During this time when Eddie was beginning to enjoy the extracurricular activities and keeping up his grades, he received an invitation to join the pre-collegiate development program through the University of Colorado-Boulder. This was a special summer program designed to prepare students to take the ACT and SAT tests as well as learn how to fill out various applications and financial forms. Eddie lived on the CU Boulder campus in the summers before his junior and senior years and took high school level courses, earning high school accreditations. "Eventually, I earned enough credits through my first three years of high school that, when I entered my senior year, I was shy of only two classes to graduate."

During his senior year in high school, the INROADS program presented an opportunity for Eddie to learn resume and interviewing skills. This program gives high-performing students career immersion (internships) for those interested in technology, engineering, and math. As part of this program, seminars were held on weekends, and upon completion of those sessions, each participant was interviewed for an internship with a company. This program arranged for Eddie to interview with Mobil Exploration and Producing U. S., Inc., one of the largest integrated oil companies in the world. He quickly found himself working with petroleum engineers each summer and learning valuable skills while he attended college. And, although he'd planned to attend the University of Colorado-Boulder, another school, Iowa State, came into focus for Eddie. Due to his strong academic background, Eddie qualified for and was awarded a full-ride scholarship to attend Iowa State, where he earned a bachelor's degree in business administration.

Eddie & friend at Iowa State, 1992
~ courtesy of Eddie Gomez

Going Out into the World

The college years at Iowa State went fast, and Eddie decided against Mobil Oil after graduation. "I came back to Colorado to work in my home state. I wanted to find my calling deciding to do assembly work at Storage Technology Corporation. I ended up helping production for new products teaching myself how to work with databases and write code for this data."

Storage Technology made the decision to move manufacturing to Puerto Rico, and Eddie had to find another position within the company. His parents at the time developed health issues, so he took a year hiatus and cared for them. This is when he met future friend and business partner, Ryan Fullmer.

Eddie Today

A unique partnership was established and Eddie provides such a wide range of talent and skills that it is difficult to list them all. In addition to creating the very successful *Healing Our Heroes* non-profit fund to treat the signature wounded veterans, he contacts, coordinates and maintains a number of generous grants and donor relationships. The funds have resulted in many civilian and military patients being treated and healed to date. Eddie does not carry one single title because his tasks and talents are so diverse that it would be difficult to list them all.

Whether it is his creating and managing the business plan and its program elements; or his scheduling, briefing and monitoring the many veterans in close coordination with Dr. Julie Stapleton; or his maintaining an effective liaison with testing, technical, and other HBOT initiatives, Eddie makes it all look easy. His quiet, low key approach is rarely noticed as he moves from one essential activity to another. His desk is never empty as papers, and projects are piled on top of each other with Eddie barely visible behind it all. Yet as many as 30 patients may be treated each day. Eddie's master schedule is a masterpiece of management and method when one observes the quiet, easy flow of people who enter and exit the clinic. This orderly scene is a true credit to his remarkable integration and coordination skills as he directs the difficult business of HBOT healing. It all happens daily, and Eddie makes it happen. A more descriptive title for Eddie and his many responsibilities might better be titled: "function junction…"

CHAPTER 4

Ryan and Eddie – The Miracle Workers Come Together

A Chance Meeting ... In Eddie's Words

As Eddie remembers what happened, he was in a Boulder restaurant, sitting at the bar and waiting for friends to arrive. Seated next to him was a friendly sort of guy, and they started talking. "We engaged in small talk and he offered to buy me a drink. At the time I was suffering from peptic ulcers and so couldn't drink alcohol. I didn't want to be rude and told him what was going on with my ulcers."

Ryan continued the conversation and told Eddie about his association with hyperbaric oxygen therapy and how it can heal ulcers. Eddie responded, "I am going in for surgery in a few weeks to have my ulcer fixed. That therapy sounds nice, though." Then he changed the subject. Eddie wasn't ready to try some hocus-pocus treatment that he had never heard about before.

He and Ryan stayed in touch, and after about two weeks, Ryan convinced Eddie to try hyperbaric oxygen therapy. Eddie went through ten treatments and his ulcers started to heal. Best of all, he discovered that he no longer needed surgery.

Eddie wondered why this *strange* treatment wasn't well known and more widely available. As his condition improved, Eddie began to volunteer at Ryan's clinic and helped him with his first "open house." As Eddie vividly remembers, "We had about forty people come to the open house, and I saw in those potential patients the need for this treatment. The people who came realized that this could be a life-changing treatment but the expense was too costly."

Ryan had previously tried to start a nonprofit while he was working for Dr. Ken Stoller in Albuquerque and was hoping to start up a nonprofit at his Boulder clinic. "When Ryan told me what he wanted to do, that really got my wheels spinning …"

Eddie began to research nonprofits. He learned the basic steps to start a nonprofit company, as well as the need to fill out the daunting twenty-four-page application. Then, he sat down with Ryan to find out just how serious he was about starting a 501c3 nonprofit organization. Eddie describes vividly, "I learned that they recommend a lawyer fill out the application … of course that was after I had already filled out the twenty-four pages."

Ryan agreed that they should move forward. On November 18, 2008, IRS Form 1023, which had been researched and filled out by Eddie, was certified through their lawyer and filed with the IRS. There was a nine-month backlog on the filing for "Recognition of Exemption" applications. In a lot of cases, applications get sent back for revision. Eddie remembers the call he received from the IRS agent, "She only had three revisions. Three! I was surprised because they usually come back with a lot more questions or requests for further information and corrections. I had them faxed to her by the next afternoon and then we waited. A few days later our fate was official … Rocky Mountain Hyperbaric Association for Brain Injuries had been granted nonprofit status. *Alright!* I thought. *Umm … err … but what have I gotten myself into?*"

A Chance Meeting … In Ryan's Words

While having a beer one day, Ryan met a young man who sat down next to him and ordered milk. Ryan, the always friendly "Mister Nice Guy," offered to buy him something stronger. As Ryan recalls, "He said, no. He told me he was going to have surgery soon on his ulcers. He seemed like a really nice guy, and we eventually segued into talking about his health issues and the chamber I was operating. After hearing his story, I related to him that I could make a big impact on his ulcers in two weeks using hyperbaric therapy, but he would have to make an appointment with the medical director, Dr. Stapleton. I gave him her number and mine and left."

A few days later, Eddie Gomez called Ryan and said that he was going to take his advice. He had seen Dr. Stapleton and procured the prescription and wanted to start treatment immediately. Within two weeks, Eddie's ulcers were much improved. He said that he was starting to enjoy his mother's cooking again. Ryan reflects on this timely meeting, "It must have impacted him, because he has turned out to be my business partner, a cofounder of our separate 501c3 Rocky Mountain Hyperbaric Association for Brain Injuries, and my good friend."

After joining forces with Ryan on a volunteer basis, Eddie began putting together a plan for the business to move forward. He produced brochures, designed a website, ran errands, and took over the administration and financial duties of both the private and nonprofit sides of the business. Additionally, Eddie began holding fundraisers and structuring the professional operation in their "broom closet" clinic as they began to seek more patients.

Ryan states that, "In the beginning, Eddie set up an informational seminar for people who were interested in the possibility of receiving hyperbaric treatment. It turned out to be a big success with about twenty-five people showing up. But after talking with all of these people we realized that most could not afford this treatment even if the price was reasonable."

It was obvious to Ryan and Eddie that the market was there, but cost prohibitive, even at Ryan's low cost per hour, which was extremely low compared to the hourly cost of HBOT therapy at local hospitals. Additionally, those hospitals did not treat veterans with TBI or PTSD in parallel, as Ryan and Eddie's clinic would eventually do. In those first days as partners, they also lacked a business plan and the advertising needed to promote this unique practice.

Eddie knew about Ryan's quest to start a nonprofit in Gunnison to help the young boy who had almost drowned in the canal. Eddie suggested that they start a 501c3 nonprofit organization and then quickly began researching to file for legal status with the IRS. Within a few months, they established the clinic's nonprofit association.

The clinic space at Charlie Hansen's manufacturing facility, which was adequate enough to hold three chambers, began to support this fledgling business. Ryan now had everything he needed and did one treatment at a time with one chamber, beginning that first year. The next year found him adding another chamber. Finally, in the third year of operation, Ryan bought his first lay-down style Sechrist hyperbaric chamber. Three chambers in a limited space brought new challenges for this growing small business. "The logistics of three chambers stuffed into this quite small space was getting pretty awkward ... and I felt we were wearing out our welcome!"

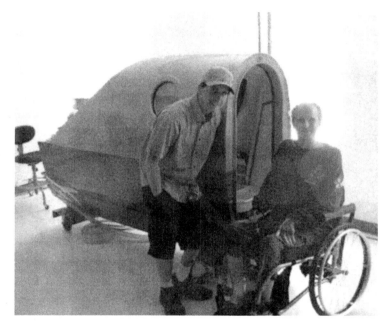

Ryan Fullmer and Charlie Hansen in front of Charlie's chamber
~ courtesy of Ryan Fullmer

Challenges

The private clinic, Rocky Mountain Hyperbaric Institute, with its nonprofit Rocky Mountain Hyperbaric Association for Brain Injuries, had now begun its initial journey into the world of HBOT. After receiving the nonprofit status, Eddie and Ryan slowly began to build their clientele and prove that this therapy actually works.

One evening shortly afterwards, Eddie and Ryan were at the clinic, stripping the wax from the floors so they could re-wax them the following day. Eddie decided to go home, because there was not much he could do to help in the process. "I went to bed and at about 11:00 p.m. I got up to use the bathroom. Something wasn't right. My whole right side was numb, and I was having a hard time standing up. *Oh my!* I thought. *I am having a stroke!* I knew the signs. Stroke runs in my family. I called downstairs to my family and scooted myself down the stairs. I waited at the bottom, which felt like forever. By then, I had lost all movement and feeling in my right side. I slumped and waited … to be saved or to die."

Eddie had suffered a hemorrhagic stroke, a brain bleed due to idiopathic high blood pressure. For the next six days, the doctors struggled to curtail his high blood pressure, which put him at risk for another brain bleed. "So, here I was, in the hospital suffering from one of the conditions for which I had been raising money to treat people with HBOT. I was extremely fortunate to have been involved with Ryan, the clinic, and our medical director, Dr. Julie Stapleton. After seven days in intensive care and two days of recovery, I began inpatient physical and occupational therapy. Once the routine was set up, I was allowed to leave the hospital one time a day to receive hyperbaric oxygen treatments with Ryan at the clinic."

After two hyperbaric treatments, Eddie began feeling a tingling sensation on his right side and started to feel his "brain fog" lift. He soon began the laborious task of relearning to walk. Recovering quite rapidly, he was released from inpatient therapy and began outpatient status as he continued to receive hyperbaric therapy.

Small miracles first happen in one's mind. Outpatient therapy was conducted at Mapleton Rehabilitation Center in Boulder, Colorado. This center was formerly Memorial Hospital, where Eddie had begun his life. He was now beginning another new life. "I had come full circle!"

By the end of three months of physical rehabilitation and HBOT, Eddie had recovered enough to be able to drive. "I knew at that point that my life's mission would be to give as many people as I could the chance to recover in the same way I had been able to do. I decided to become a certified hyperbaric technologist. To do this, I had to become a certified nurse's aide, attend class in San

Antonio, Texas, complete 450 hours of internship, and sit for the National Board exam. Easy, right? Well, I am proud to say that I was able to accomplish all of this before I was a year post-stroke. I met my goal. And the rest, as they say, is history!"

ROCKY MOUNTAIN
HYPERBARIC
ASSOCIATION
For Brain Injuries

Logo created by Welch Creative Group, Denver, CO

The Hardest Part – Funding

Unfortunately, not everyone understands ideas or mission when a clinic is newly established. Fundraising, in and of itself, is a whole different world. It can become a very complicated world with many different ways to obtain seed money. In recapturing the startup days, Eddie tells of their first break, "We were very fortunate in the beginning to have a resident of Boulder who had been treated in the clinic. Along with her husband, they provided the initial funding that the nonprofit needed. The couple's generous donation allowed us to provide scholarships to individuals to help them pay for their treatments. This donor couple continues to be a strong financial support for the nonprofit."

After this initial donation, Eddie then tried his hand at grant writing. He began to acquire an appreciation for the very daunting task. After twelve rejections, the clinic's nonprofit finally received a $7,500 grant to help provide treatments to wounded veterans. It was during this same time that Eddie had organized an informational seminar at the Caritas Center in Boulder to both discuss HBOT and create awareness of the nonprofit program.

As Eddie recalls, "I had mentioned our Caritas Center meeting on Facebook, and through this post, I met a Marine, Bob Alvarez, counselor in the Warrior Transition Unit from Fort Carson, Colorado. He had previously helped coordinate some of the brain-injured veterans, who'd returned from Iraq and Afghanistan, so that they could receive HBOT in other clinics. Bob was hoping to bring an interested retired Marine Colonel to this seminar." Colonel Robert L. Fischer attended with Alvarez but remained skeptical until he saw firsthand the success these treatments produced. Fischer began to frequent the clinic and observed what we were doing.

After that first meeting in 2010, Fischer became totally committed to their fledgling program, especially after he interviewed a few of the first veterans that Ryan and Eddie had treated. He became convinced that HBOT was safe, effective, and the future in treating TBIs. Shortly thereafter, he began his own personal search throughout all of his contacts to help find funding for the Rocky Mountain Hyperbaric Association for Brain Injuries, which is when he enlisted the help of his friend, Grady Birdsong.

Colonel Bob Fischer and Grady Birdsong had a number of veteran organization contacts in Denver and nationwide. They started presenting the Rocky Mountain HBOT program to others and gained early monetary support for the nonprofit fund that Eddie had established in 2008. As part of the nonprofit organization, Ryan and Eddie subsequently created a special fund and named it the "Healing Our Heroes" program to help veterans with treatments and housing costs while undergoing treatment.

The largest Marine Corps luncheon group in the Denver area, "Cooper's Troopers," the American Legion, and the VFW raised the first veteran-donated funds. Other donations trickled in from various social, military, and government groups. Grady went to work creating short videos of the veterans' dilemma and the healing process taking place at the clinic. After making videos for the clinic website, Bob and Grady made numerous presentations using Grady's short videos.

In seeking donors, Bob also contacted one of his acquaintances, Peter Haas, the President of Marine Corps-Law Foundation, whose headquarters is located in New York City. Mr. Haas replied that his Law Foundation charter did not allow financial support of such programs, but rather their funds were used to defend Marines and their families who had serious legal problems.

However, once he read their material, Pete contacted another friend who was literally on his deathbed. When he showed his friend Grady's video, the dying man wrote a check to Eddie's "Healing Our Heroes" 501c3 fund for $50,000. This was the beginning of bigger things to come.

Shortly after that, the biggest and most important fund benefactor joined the veteran advocates supporting the RM Healing Our Heroes non-profit fund. Marty Hoffman, the former Secretary of the Army under President Gerald Ford, was one of the first to support the fledgling HBOT program of Dr. Paul Harch in New Orleans, along with Caleb Gates, Vice President of Denver Investments, who helped raise the first funds to treat and heal the first veterans who suffered from TBI and PTSD. Later, Marty introduced Bob Fischer and Caleb and recommended that Fischer show him what the Rocky Mountain HBOT Clinic was also doing to heal war veterans. As Caleb recalls "it was love at first sight" and it turned out to be the most significant event in the early years of the program. Even though his first meeting with Ryan and Eddie took place in the Boulder "broom-closet" clinic, Caleb quickly saw the exceptional results of Ryan's HBOT therapy and also heard it confirmed by the healed veterans themselves. Today, his large fund-raising galas continue to support and ensure the future success of the Rocky Mountain Hyperbaric Clinic that has treated and healed more than 300 veterans.

CHAPTER 5

Dr. Julie Stapleton – In Her Own Words

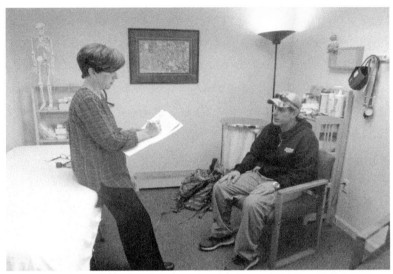

Dr. Julie Stapleton examining a Veteran suffering from TBI
~ courtesy of Grady Birdsong

B orn in Detroit, Michigan, and growing up in a family of five children, Dr. Julie became interested in healthcare when her dear mother went back to school to become a nurse, which was motivated by the need to pay for Julie's older brother's medical school tuition. As Dr. Stapleton describes the beginning of her life's journey, "I went off to school to become an occupational therapist, but because I enjoyed and excelled in my first years of college, and combined with my mother's influence, I instead opted to pursue the practice of medicine."

Nearing the end of medical school, Dr. Julie discovered Physical Medicine and Rehabilitation (PM&R) as a specialty. This brought her full circle back to her original interest in occupational therapy, since PM&R is a specialty whose philosophy is based on providing patients the opportunity to recover from injuries and regain their maximal functional capacity. PM&R physicians are also known as physiatrists and treat a wide variety of medical conditions affecting the brain, spinal cord, nerves, bones, joints, ligaments, muscles, and tendons, focusing on the entire body in hard-to-diagnose problems.

"When I got further into rehabilitation as a specialty, I became intrigued by the opportunity to rehabilitate individuals with TBIs. Rehabilitation of such a devastating injury allows me, as the physician, to work closely with my patients and their families over an extended period of time, watching their initial survival from a catastrophic injury, through recovery, through the process of regaining functions, and eventually to the capacity to get back their life ... that is what I find most rewarding, especially those injured combat veterans I now interview and prescribe treatment for at the Rocky Mountain Hyperbaric clinic in Louisville, Colorado."

Hyperbaric Oxygen Treatment

Dr. Julie Stapleton's introduction to hyperbaric oxygen treatment came through her established medical practice in Boulder, Colorado. About ten years ago, through her patient Charlie Hansen, she became intimately involved with HBOT treatment. When Charlie asked her about HBOT treatments, Dr. Julie told him what she often told others, "I think it is safe, since it is dealing with oxygen under pressure, yet there is no concrete evidence that it works. But there is no harm in trying."

After some research on chambers, Charlie bought his own chamber and installed it in the back of his Boulder business. His intent was to train some of his family members and key employees to administer the treatments. Once he got the chamber, he became quickly overwhelmed, realizing he needed more help. That's when he began to look for a trained individual to properly operate the chamber, and that trained technician turned out to be Ryan.

As Ryan set up the chamber to treat Charlie, the two of them began conversing about a mutual dream, that of starting a hyperbaric oxygen clinic. They discussed plans to offer oxygen treatment to other patients, but soon realized, per the FDA regulations, that they would need a medical doctor to screen each patient and write a prescription for the HBOT technician to operate the chamber.

Quickly, they approached Dr. Julie. The beginning of the "miracle team" had begun. Things were looking up.

As Dr. Julie recalls, "Since I was the one who wrote the prescriptions for Charlie, I also referred a couple of my other patients to them so that I could learn a little more about the treatment and witness the results firsthand." Agreeing to do so for Charlie and bringing in a few of her other patients, Dr. Julie told Ryan, "Okay, fine. I am in as medical director on one condition … it has to work! If it doesn't work, I am out of here!"

As Dr. Julie now confirms, "It has worked and has continued to be an amazingly rewarding experience."

Admittance – The Screening Process

Dr. Julie has become quite familiar with the technology and politics of hyperbaric oxygen treatment. She tells us that the FDA has determined HBOT to be safe and effective but has limited their approval to fifteen specific diagnoses, including decompression sickness, radiation tissue damage, diabetic wound healing, and carbon monoxide poisoning.

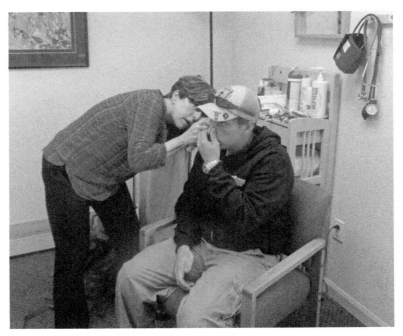

Dr. Julie examining Veteran patient
~ courtesy of Grady Birdsong

HBOT is not approved for TBIs or strokes, yet amazingly, the drug employed in HBOT is one of the most common in nature . . . oxygen. As Dr. Julie explains it, "There is no single drug in the United States that is FDA approved to treat TBI. Every single drug I use in my patient population is borrowed from another diagnosis, which is perfectly legitimate, legal, ethical, and considered safe. It is done all the time across the board in multiple specialties. I write a prescription for oxygen under pressure, then I give the criteria for how much oxygen, how deep the pressure, how frequent the treatments, and the total number of treatments that should ideally be aimed for. The only thing that the FDA controls in this situation is the oxygen." She further elaborates, "While it is safe and ethical to use drugs off-label, it is not required that insurance companies pay for things off-label, and so they most often do not."

Dr. Julie first screens patients for indications or conditions that might make hyperbaric unsafe. Longtime smokers, who have emphysema or potential lung cancer, or anyone who has had recent chest surgery, are considered patients who may not handle the pressure that is administered in the hyperbaric chamber. The pressure depth administered in the treatment of TBI and/or PTSD is 1.5 ATA, which is the same as 1.5 times the amount of pressure at sea level, which is roughly the equivalent of going 17 feet below sea level here in mile-high Colorado.

Dr. Julie's screening approach is simple. "My first goal is to make sure the patient doesn't have any contraindications. My second is to start the process of educating them on how to manage their ears in the chamber while undergoing compression. My third is to provide them with education and reassurance that it is safe. Finally, my goal is to answer all of their questions and again reassure them that the treatments are safe and comfortable."

Treating Veterans

Dr. Julie observes that the veterans who come to her are somewhat different in presentation compared to her civilian population—the traumatic events that have led to their concussive injuries (likely caused by either a blast injury or motor vehicle accident, which in turn may have been caused by the IED blast) are also very likely to be associated with significant combat experience with a possible associated secondary diagnosis of PTSD. Also, it is possible that these individuals have had multiple prior concussive events, which makes their recovery even more challenging.

~ courtesy of Grady Birdsong

In describing the phenomenon of treating both post-concussion syndrome (PCS), a minor form of TBI, and PTSD, Dr. Julie explains that "fear is a big part of the traumatic event." She takes it a step further and adds, "I believe that the fear component that these individuals are experiencing at the time of their injury has a direct impact on their symptoms and recovery. For example, we know for a fact that people who have had past traumatic histories are at greater risk for having more severe and prolonged consequences of a mild brain injury. It is possible these individuals are at a heightened fear response because of PTSD in the field. These veterans may be more likely to have exaggerated and/or prolonged sequelae [after effect or secondary result] of their injury. It is a true neurologic predisposition that makes them more vulnerable to these injuries."

According to Dr. Julie, a component of that PTSD can be seen "in the hyper-vigilance these individuals experience that manifests as anxiety, insomnia, and irritability—also noticeable after a brain injury. It is the association of TBI and PTSD that is very difficult to tease apart in the military patient population. It presents as a decreased capacity to manage and respond to anxiety, or to anxiety-provoking situations, compared to their pre-injury or pre-combat, or pre-post-traumatic state. Significant effort is being made in both the military and civilian medical communities to help sort out and optimally treat both these coexisting conditions."

When asked if there is a connection between individuals' physical and psychological wounds, she responds, "I don't like it when people try to make that distinction between neurological versus psychological, or organic versus psychological. Do they think that emotions and psychological well-being come from the heart? No, these can be manifestations of changes in the normal functioning of the brain."

Experiences

While some experts claim that TBI is a physical injury and HBOT provides the physical healing therapy for TBI, yet PTSD is a psychological injury and requires a much different therapy, she quickly responds, "Not true! They can both be organic brain injuries!" Dr. Julie believes PTSD can respond to HBOT in the chamber, and the results are optimal if supported by complimentary trauma counseling during the weeks of hyperbaric therapy.

Dr. Julie honestly believes that she doesn't have all the answers for HBOT and whether it works for PTSD as well as it does for TBI. However, she bases her opinion of treating PTSD on the experience of having treated a civilian patient who came to her with disabling PTSD symptoms without ever receiving any physical injuries.

Her patient had witnessed a horrific motor vehicle accident, which had only narrowly missed her own vehicle when she and her young children were stopped at a red light. Dr. Stapleton explains, "That woman experienced a singular moment in time, which I describe as the 'oh shit' moment, where an individual experiences the fear of death or of serious injury to themselves or their loved one, and they are completely hopeless or helpless to do anything about it. That moment triggered in this patient the fight or flight reaction, which then persisted, manifesting as post-traumatic stress symptomatology for well over a year. She couldn't sleep, was hyper-vigilant, depressed, extremely anxious, couldn't handle stress, couldn't focus her attention, couldn't problem solve, or make decisions ... classic textbook symptoms of PTSD. I prescribed forty sessions of HBOT and her symptoms resolved."

As justification, Dr. Julie reminds us that, "This story is an example of the definition of what is required to have PTSD and how it changes one's psyche. Traumatic experience can change how the brain operates, and there is neuro-scientific evidence of that. It is not simply psychological, it is neuro-physiological."

She makes another important distinction about the coexistence of PTSD and concussion. PTSD doesn't occur in other events, such as sporting events where the individual has chosen to participate in a risky activity, is aware of and assumes that risk. These individuals may well sustain concussions and are vulnerable to multiple concussive injuries, for example, the types of head injuries that are now being recognized in the NFL, but without the coexisting condition of PTSD. "You make that choice and you know you are going to be in a violent, potentially risky sport. You do not experience that overwhelming fear and are not at risk for triggering the fight or flight reaction. That outcome is different than somebody who gets in a car accident or, worse yet, is the victim of an assault or sustains an injury under combat conditions."

The Healing Timeline

Dr. Julie points out that when a patient is in the chamber for the prescribed treatments, there often comes an obvious point when their energy, memory, and cognitive faculties start to come back and a new sense of wellness comes into play. At that critical point, the veteran may experience a tremendous psychological boost or a big emotional letdown. It is different for each person. "And so it does help to have a counselor on the team, like Pepe Ramirez, a combat veteran who has been there. These vets, if they do not have good coping skills or good communication skills, and if alcohol and drugs are a factor, may go back down a path of regression. Without the right kind of support system when they leave our clinic, it will look like HBOT didn't help the PTSD."

Dr. Julie also continues to learn from past experience and her patients. She further emphasizes, "You have to look at the prior history of the patient. How did they function pre-injury? What challenges did they endure? Do they have good coping skills? Do they have good family support?" She contends that if these skills are not there to begin with, then it really makes the rehabilitation process all that much harder for the patient who is dealing with the dual complexity of PTSD and TBI. At the Rocky Mountain Hyperbaric Institute, veterans benefit from the integration of both kinds of treatment.

Dr. Julie further stresses that HBOT doesn't work for everybody with TBI and/or PTSD, and forty hours in a chamber is not the magic number of treatments for everyone. It may be just the minimal starting point. Experiences with severely injured veterans who suffer TBI and PTSD confirm this.

Past, Present and Future

When asked why hyperbaric oxygen treatment is not an accepted practice in the medical community, Dr. Julie thoughtfully replies, "Because nobody knows about it yet ... they look at old literature, which does not show the success and strides we have made more recently, or become understandably confused and discouraged by the research carried out by the Department of Defense. However, due to the pioneering work of Dr. Richard Neubauer and Dr. Paul Harch, there are numerous articles, books, and well-designed studies that demonstrate the benefit of HBOT in the treatment of mild TBIs.

Dr. Harch has been involved in educating and documenting these benefits for several decades. Despite that, the benefits of hyperbaric therapy are often considered as *anecdotal*, which implies favorable human interest stories suggesting good outcome rather than solid scientific research. It has been extremely challenging to perform controlled studies in HBOT like you can with medication. It is hard to design a placebo, since it is hard to fake a hyperbaric oxygen treatment. Despite that challenge, progress is being made, and evidence is mounting that HBOT is a safe and effective addition to a comprehensive multi-disciplinary treatment approach."

What Keeps Dr. Julie Stapleton in the HBOT arena?

"Hyperbaric therapy actually treats the problem rather than just putting on a Band-Aid. In my treatment of patients with TBI, I first treat sleep and pain, and then I compensate for cognitive and emotional issues, mostly through medication, education, and therapy. But hyperbaric oxygen treatment really does heal the brain! Long ago, I used to say that acupuncture likely works but has no good evidence-based science to back it up. Now I am an acupuncturist and I have no doubt that it works. I also have no doubt that hyperbaric therapy will soon be readily recognized as a mainstream treatment option, as acupuncture has become."

Dr. Julie is still very much in the game because of all the success she witnesses. She revels in the stories about veterans who tell their buddies that they should come to the clinic and try it out. One of the first questions she asks a new veteran who is checking in is, "How did you hear about us?" Their response is often, "Through a friend … another veteran who you helped!"

What the Veterans Think about Dr. Julie and HBOT

What is most obvious about Dr. Julie's practice is reflected in the comments of the many veterans she has treated at the clinic (and the civilian clientele as well). She is held in very high regard, and that respect is reflected in their comments about her.

One veteran responded, "Dr. Julie. She really cares for us. You can tell by the way she talks to you and the nice way she gets your medical information that she really likes us. She puts us at ease when we first meet her, and her smile and demeanor give you a sense of confidence that she will really help heal us and make our lives so much better."

 Dr. Julie Stapleton of Boulder, Colorado, joined the Rocky Mountain Hyperbaric team in 2007. She commits time and energy from her established practice of physiatry to function as the medical director for the Rocky Mountain Hyperbaric Institute. Dr. Stapleton graduated from University of Michigan Medical School in 1985 and has been in practice for thirty-one years. She completed a residency in the University of Michigan Health System in Ann Arbor, Michigan. As a clinical professor while at University of Michigan, she administered the TBI Day Treatment Program as its medical director. She currently practices at Julie A. Stapleton, M.D., and is affiliated with Boulder Community Health. Dr. Stapleton is board certified in Physical Medicine and Rehabilitation and a Certified Hyperbaric Physician.

CHAPTER 6

Sergeant Major Pepe A. Ramirez, USMC (Ret) – In His Own Words

Pepe Ramirez, Counselor
~ courtesy of Eddie Gomez, President, RMHBOT Assoc.

How did a retired Sergeant Major of Marines come to this exceptional team of miracle workers? What could he offer to wounded veterans with TBIs and PTSD? And how did the Rocky Mountain Hyperbaric Institute become the first integrated TBI-PTSD treatment program in the nation when VA hospitals and treatment centers still separate the two conditions?

Returning to Denver, Colorado, after retiring from the Marine Corps, Pepe Ramirez embarked upon a field of study that would alter his future life. Because he still loved the men he'd led in the past, Pepe immediately enrolled in the University of Denver and completed a Master's degree in Social Work, a course that contained both Existential Therapy and Cognitive Behavior Therapy. He chose this field of study due to his past experiences as a leader in the Marine Corps. He felt strongly that he could help other veterans with their transitions to the civilian world.

Recently, an anonymous combat veteran patient of his told it so well: "This Sergeant Major has been there. He has experienced three combat tours in Iraq. When he talks to you the connection is immediate. He knows exactly how each of us feels and why. He does not type away on a computer and barely notice you as most counselors do. There is real bonding and immediate trust in him that he can really help us ..."

Pepe Ramirez admirably achieved the highest and most respected enlisted rank in the United States Marine Corps, yet that is not all that can be said about him. The real measure of Ramirez is how he managed to succeed when he was faced with more obstacles than most people dream about in a lifetime.

After being born in the Philippines, his family moved to the south side of Chicago, Illinois, when he was nine years old. Every street had its own gang: Italian, Irish, African American, Hispanic, etc. Pepe quickly learned to fight to survive. Referred to as the "chink," he received his first "basic training" on those streets and learned the fine art of survival. Pepe recalls, "The fights were their special way, their ritual of welcoming me to their neighborhoods and sizing me up."

Later Pepe's family moved to Houston, Texas, where he graduated from Klein Forest High School. Throughout his growing up, he was continually inspired by his own family members who had served our country's military. The year 1988 found him solidly immersed in 3rd Recruit Training Battalion, Marine Corps Recruit Depot-San Diego, beginning his brilliant and successful career with the Marines. He was meritoriously promoted to Lance Corporal upon graduation and selected to attend machinist school at the Aberdeen Proving Ground in Maryland. After completing his schooling, it was off to 3rd Force Service Support Group in Okinawa, Japan, where he quickly and meritoriously picked up the rank of Corporal. Finding his way into the non-commissioned officers' school, he quickly became a meritorious Sergeant with less than two years in the Corps and found himself responsible for Marines as a Platoon Sergeant.

As the Persian Gulf War began, Pepe received orders to the Philippines in 1991. However, there would be no combat tour just yet for the Sergeant. He first attended jungle environment survival training and learned to survive in a hot and hostile jungle with a bag of rice, a machete, and a small first aid kit.

After jungle school, he reported to Quantico, Virginia, becoming a Marine combat instructor in water survival. As he describes, "It was the most challenging course I ever attended. I started the training at 135 pounds and wound up weighing 105 pounds by the end of the training. Only seven of the fifty starting students graduated." This training helped him later, when he was able to save a young girl from drowning and was subsequently awarded the Navy Achievement Medal.

Learning Leadership and Making Marines

In 1994, Pepe served in the Marine Air Logistics Squadron 36 at Futema Air Station, in Okinawa, Japan, and became a Staff Sergeant. It was then that he received orders to return to Marine Corps Recruit Depot-San Diego, California, to become a Drill Instructor, one of the most coveted of assignments in the Corps.

The beginning of leadership is well established on the drill field. It is the most important job and the essence of the Marine culture in forming the finest combat organization in the entire world. While there at the depot, Pepe again attained a meritorious promotion to Gunnery Sergeant. When asked what one thing he had learned and now uses, Pepe relates, "I realized that my leadership skills had increased exponentially as I learned which of the "hot buttons" to push to create a positive behavior in my recruits. Drill Instructor duty helped me understand my own limits as well as the human spirit. It would serve me well for the rest of my Marine career and life."

Letting no grass grow under his churning feet, in 1999, Pepe's next duty station became Camp Pendleton, California, with Ordnance Maintenance Company, 1st Maintenance Battalion, and 1st Force Service Support Group. As a Gunnery Sergeant, he had many jobs, including Director of the Corporal's Course and Chief of Intelligence and Training (S-2 and S-3). He also found the time to earn a Black Belt Instructor certification in the Marine Corps Martial Arts Program. Additionally, Pepe enrolled in night school, which led to a bachelor's degree in psychology. By April 2002, he became a 1st Sergeant.

Senior Drill Instructor S/Sgt Ramirez, MCRD-SD
~ courtesy of Pepe Ramirez

Combat Tours

By 2003, Pepe became attached to Regimental Combat Team 7, an infantry unit in Kuwait poised to strike in Iraq. His unit, Combat Service Support Company-117, engaged in Operation Iraqi Freedom and the push up to Baghdad. By October 2003, he returned home only to learn that he would re-deploy again to Iraq in four months. He returned for Operation Iraqi Freedom III despite being advised that his oldest son had been diagnosed with cancer. Pepe's family and his son mentally supported another third assignment to Operation Iraqi Freedom III, one of the most intense assignments yet for the Marines. Pepe witnessed TBIs and post-traumatic stress firsthand. There are not textbooks for the circumstances and realities of duty in three tours of combat.

A happy birthday to son, Marcos
~ courtesy of Sgt Major Pepe Ramirez, USMC

Achievement in Tough Times

Pepe was promoted to the most admired and sought after enlisted rank the United States Marine Corps bestows: Sergeant Major, USMC. His family— wife, Rose, and their six children—attended his promotion ceremony to that distinguished rank in March 2007. Sadly his oldest son, Marco, succumbed to cancer two weeks later.

The Marine Corps did not miss the opportunity to utilize Pepe's expertise as he began winding down his career. Reporting to Marine Forces Reserve Headquarters in New Orleans, Louisiana, Pepe established and standardized the Combat and Operations Stress Control Program (COSC). This program focused on families who were experiencing deployment or pre-deployment to combat duty. It encompassed training and certifying all coordinators at all of the 180 reserve unit sites throughout the United States. Under Pepe this program teamed with all Family Readiness Officers, Substance Abuse Officers, Mental Health Providers, and the Health Assessment programs. These were his duties as the Regional Training Coordinator for the COSC.

Sergeant Major Pepe Ramirez, USMC (Ret)
~ courtesy of Grady Birdsong, Veteran Advocate

After retiring and obtaining his master's degree, Pepe established Combat Vets Helping Combat Vets and was contacted by Eddie Gomez to talk about synergies between the two organizations. Ryan and Eddie saw the potential he had created in his own consulting business, which focused on helping both combat veterans and their families suffering from post-traumatic stress. This was the point at which Ryan and Eddie recognized that they needed someone of Pepe's stature and military bearing who could deal with the veterans who were coming through the program.

This affiliation happened quickly, and they began the integrated approach to treat both maladies, TBI and PTSD, in parallel. This team, the Rocky Mountain Hyperbaric Association for Brain Injuries and the Combat Vets Helping Combat Vets, is the only known integrated TBI-PTSD program in the country, which now has a documented and successful track record, complimented by Sergeant Major Pepe A. Ramirez, USMC (Ret).

The logo
~ courtesy of Pepe A. Ramirez, Counselor

New paths and ways of thinking have been forged as a result of their integration of treatment for brain injuries and post-traumatic stress. Pepe explains the methods he employs in dealing with veterans: "During our hyper-baric treatment for TBI, I am right there when they open up about anger, guilt, grief, and other things that cause them consternation and withdrawal. In addition to my hands-on PTSD counseling during the TBI treatment program, I lead a physical exercise program at a local recreation center. It is another dimension of the mind, body, and spirit program that I provide for them."

This program re-establishes and increases both strength of the body as well as the clarity and coping skills of a confused mind that is starting to regain lost memory. Pepe reminds them of days past. "This is all in sync with the treat-ments that provide physical healing of the brain and body. It is easy to see that these exercises give them renewed confidence. Then the spirit usually follows if that warrior believes in a higher power. The former military esprit is certain to return when I motivate them to accept this new and healing life that hyperbaric treatment provides. In this way, I help them return to the confidence gained from graduating from recruit training."

The Sergeant Major goes to an even deeper level as he explains what happens in a way that can only be understood through personal experience. "I help them renew and reflect the coping skills they had with their combat buddies when a trusted friend had his "six" (back) and could cope with the worst conditions he's ever experienced. I have to teach them this all over again and help them out of the withdrawal and mental suffering they have regressed into since coming home. When they realize that again, they remember that unique support system."

Mind, body and spirit training
~ courtesy of Pepe A. Ramirez, Counselor

Pepe also employs EMDR (Eye Movement, Desensitization, and Reprocessing). This is a form of psychotherapy that was developed by Francine Shapiro, Ph.D. to help resolve the development of trauma-related disorders caused by distressing, traumatizing, or disturbing experiences. Pepe explains this methodology in simple terms. "Basically what happens is that it reconnects the left side of the brain hemisphere to the right hemisphere so they are working in concert. Dr. Shapiro was walking in the park one day and she was engaging in lateral eye movements, looking left, looking to the right. Once she finished her walk, the disturbing thoughts she was having disappeared. She then began studying this and developed procedures to achieve results."

Pepe expands by adding, "Remember when cadence was called in boot camp? I called cadence to my recruits: left, right, left, and platoon halt! Technically, we were all doing EMDR. When we did physical training, PT, we were calling cadence. And at those moments we were really doing that eye movement, desensitization reprocessing. This is the reason that I feel physical fitness and calling cadence is very therapeutic."

Pepe is always on the lookout to evaluate new promising PTSD therapies. And he never forgets the wife and the family in a veteran's healing process. Using the analogy of gears meshing and interlocking, an interrelationship between a veteran and his family must be adopted and used if each is to understand the cognitive process that helps heal the family system. Pepe is emphatic on this point. "Many wives reply to this gear analogy by saying, 'I now have my husband back and the kids also see the change in their father.' It is almost too good to believe."

VETERAN

SPOUSE   CHILD

Cog wheel analogy used in counseling

Hard Choices – Reflecting on the Past

Reflecting on the past, Pepe uses a harsh but honest approach to hopefully bring the veteran back to present reality. As he describes this procedure, "In those difficult cases when a veteran and family are close to divorce and the children suffer from the obvious mental condition of the veteran, I use hard questioning in my sessions to help the veteran cope with that renewal. Many veterans lose the love and support of the whole family if they have no post-traumatic stress treatment or mental help."

Pepe contends that his techniques in counseling get their attention fast. Along with the therapeutic healing, he teaches them "to be the person they once were ... always continuing to be the warrior ... but again assuming the responsibilities of being a warrior."

Pursuing a Ph.D.

Pepe Ramirez is currently working towards earning his Ph.D. in Clinical Psychology concentrating in Neuropsychology.

CHAPTER 7

Brain Neurotraumas

Brain neurotrauma is sometimes misdiagnosed and still not totally understood in the neurological world of medicine. But new methods and therapies look more promising, as neurologists and brain trauma specialists unravel the complexities of brain injury. The patients who are lucky enough to access expert care, specialized treatment, and support can now expect a healthier lifestyle afterwards, just as Ryan was able to discover in his young adult life.

Little did Ryan Fullmer realize when he first became acquainted with hyperbaric oxygen treatment that he would ultimately "pioneer" his own special HBOT program that treats and heals combat veterans suffering from not only TBI, but also PTSD, the latter being a complex, interwoven mental condition that sometimes accompanies the physical trauma, as described earlier by Dr. Julie Stapleton. Both conditions can result when a concussive blast damages the brain, like our veterans experience with IEDs, mortars, and other concussions on the battlefield.

As described earlier, these traumas have been appropriately called the signature wounds of our most recent wars. It is estimated that almost thirty percent of combat veterans come home with both of them. The latest Rand Corporation estimate from 2008 is that 330,000 have TBI of some degree. About the same number have PTSD. The DoD and the Defense and Veterans Brain Injury Center (DVBIC) reported on August 18, 2015 a total of 333,169 for all severities (see Exhibits A and B). Further predictions indicate there may be as many as 500,000 with TBI and 800,000 who suffer from PTSD.

Even though these diagnoses, TBI and PTSD, have recently gained much attention in the military and medical communities, there are still many unanswered questions. Adequate and sufficient treatment is still lacking and this is the subject of considerable debate in Congress, the medical profession, and in the military itself. There are still no clearly accepted diagnostic processes or practices recognized or employed by the medical profession, which is also yet to develop a "tried and true" proven methodology or therapy to treat these debilitating injuries. Slowly, though, it is the Ryans and Eddies of the world who are proving there is another solution.

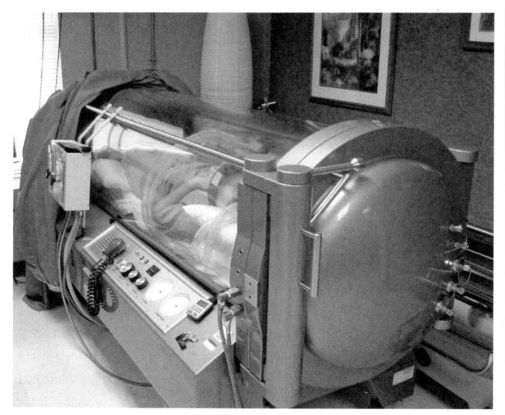

Hyperbaric chamber at Rocky Mountain Hyperbaric Institute, Louisville, Colorado
~ courtesy of Grady Birdsong, Veteran Advocate

A large majority of medical doctors question the effectiveness and reliability of hyperbaric oxygen treatment for brain injury, especially the concussive injuries young veterans now suffer. VA hospitals will not adopt the use of HBOT treatment for veterans with TBI until the FDA approves it. Instead, they prescribe dozens of medicines (off-label drugs) that are not approved by the FDA and in some cases have proven to motivate a suicidal result. There are alarming numbers of veterans committing suicide every day.

Rocky Mountain Hyperbaric Institute's record of treating and healing over 250 veterans with TBI and PTSD cannot be ignored. The final measurements and successes are not the claims of the naysayers but the validation and verification by the veterans themselves.

CHAPTER 8

Veterans Treated and Healed by HBOT – In Their Own Words

Four combat veterans of the more than 250 veterans treated at the Rocky Mountain Hyperbaric Institute have been selected for their courageous accounts as well as for their personal devotion to all veterans who suffer the signature wounds of war. All four of them suffered serious brain trauma in the Iraq-Afghan wars, at a time when the benefits of HBOT were virtually unknown. These veterans remained untreated, but over-medicated, until they were rescued by HBOT therapy. We would like to tell every veteran's story but have settled on these four combat veterans who credit their own amazing recoveries to this therapy. We can assure you that each healed veteran would tell you a similar story. They all have undergone HBOT treatment and some form of professional PTSD counseling.

Sergeant Margaux Mange, U. S. Army (Retired) Military Police – Iraq

In 2012, Charles "Pat" Smith, Department Adjutant of the American Legion of the State of Colorado, invited Colonel R. L. Fischer to meet a young woman who had recently been medically separated from the Army and who had just begun successful hyperbaric oxygen treatments with Dr. Paul Harch at his clinic in New Orleans.

At this meeting was the editor of the Legion's *Observer* Magazine, Darrel Myers, and former American Legion National Commander in 2005, Thomas Bock. The reason for the meeting was to gain a deeper understanding of this new treatment that was proving to be effective in healing concussive blast injuries suffered by our combat veterans.

**Colonel R.L. Fischer, Charles "Pat" Smith - Colorado Legion Adjutant
and 2012 American Legion National Commander, Fang A. Wong**
~ courtesy of Grady T. Birdsong, Veteran Advocate

Past Legion Commander Bock prefaced the reason for the meeting by telling Colonel Fischer, "Wait until you meet this young lady. She has quite a story to tell us." Shortly thereafter, the very attractive young woman arrived with her friend in tow, a shy lanky young man, obviously a veteran. She greeted everyone there by introducing herself and then turned to her companion to introduce him, "You all will have to speak slowly. My friend is a United States Marine." That comment broke the ice and the group truly enjoyed meeting with and listening to Margaux Mange tell her story from playing high school soccer to becoming a United States Army Military Policeman. Here is Margaux's story.

Sergeant Margaux Mange was one of the South Boulder Road clinic's first patients. Margaux grew up in Lakewood, Colorado, playing soccer throughout her childhood and teenage years. She enjoyed the sport and became a star soccer athlete at Alameda High School. Margaux admitted that the major reason she joined the Army was the promise of playing soccer for them. As she vividly described her reason for entering the military, "The recruiter assured me that I could play on the All-Army Soccer Team, and since my father was in the Army it made sense for me to join, even though my father wanted me to join the Air Force. It was the better branch for me in his eyes."

Joining the Army

Margaux began her basic training at Fort Leonard Wood, Missouri, in 2003 and was assigned the military occupational specialty (MOS) of Military Policeman. After completing basic and military police schooling, she was ordered to Kitzingen, Germany, her first duty station. Upon arriving in Germany, her soon-to-be Platoon Sergeant picked her up at the train station. He immediately leveled with her in no uncertain terms and, in effect, told her, "I know you joined the Army to play soccer, but that is not going to happen. The Army has other plans for you and you will soon be going to Baghdad."

She did get to play a few months of soccer while going through her military training at Hohensfels, Germany, in Bavaria for FTX. She won MVP honors on the team after returning from Iraq. "By the following April 2004, I was in Iraq with the 630th Military Police Company and we started going on patrols immediately. At first we patrolled all over Iraq and trained a lot of Iraqi police. Two of my missions were to Sadr City, Iraq, and it was an incredibly dangerous area with known ambushes and a lot of improvised explosive devices."

Margaux on patrol
~ courtesy of Margaux Mange, 630th MP Company

Her MP Company was also engaged in setting up protective concrete barriers around the Iraqi police stations so that insurgents could not shoot into or run speeding cars with car bombs into those facilities. They also were assigned the responsibility of escorting senior commanders in and out of the Green Zone, the Army's Supreme Command Headquarters in Baghdad. "I remember some of the firefights, the mortar attacks, and some of the deadly IED concussions we experienced. I survived that first year of my deployment because the explosive devices were poorly made and we could recognize a wire or dead dog [wired with explosives], tipping us off that there was an explosive device nearby."

Margaux Mange in Iraq
~ courtesy of Margaux Mange, U. S. Army (Ret)

When her tour of duty came to an end in April 2005, she returned to Germany and resumed her normal MP duties. Margaux reveals that she "hated my job since it involved writing and issuing tickets. I confess; I only gave out a couple of them in my entire military police career."

After being in Germany for a while, Margaux hoped that she could again try out for the soccer team. She was disappointed when she was sent to train for her next OIF tour as a gunner instead of a driver.

Second Time Around

Margaux had been a driver on her first tour, but this time she was assigned as a turret gunner and would man a machine gun and also a grenade launcher. Like her first tour, she was sent back to Baghdad. "There I manned a M240B machine gun fired from a vehicular mount in the Humvee turret. Also mounted in the turret was an Mk 19 grenade launcher, which was useless in the city due to close quarters. We conducted the same missions as on my first tour, focusing on police training at and around their stations for long hours of the day. We were charged with accounting for all captured weapons and watching for typical corruption when they sold their arms or were paid off to not get engaged. On this tour, the IEDs became more sophisticated and some of them had chemicals in them that would burn through a body part, like white phosphorous. The insurgents also had acquired rocket-propelled grenade launchers [RPGs] that were quite deadly."

Margaux received a massive concussion while on patrol in her Humvee on December 4, 2006. The explosive device was detonated approximately ten meters away from her Humvee. As she recalls, "My head was violently thrust backward and hit the turret. I was unconscious for about thirty to sixty seconds. I really didn't think anything of it. When I came to, I continued to stay in the *kill zone* for an hour, waiting for the recovery vehicle. When I first climbed out of the turret, I was seeing stars and then I was down for three days. I still didn't think about it and went back to my regular job. At this time I was only concerned about my other teammates."

After this incident, Margaux continued going out on patrols, doing her regular job as a gunner and not thinking that anything had happened to her. She was more concerned about the others on the team than herself. Their combat patrols became much more dangerous. As she describes, "It didn't affect me badly until March 2007, when my best friend, Ashly, and two others were hit with a fifty-pound explosive device. I was riding on truck #5 and they were in the truck right behind me, in Truck #6. When the IED detonated, it flipped their truck completely upside down and killed all of them instantly. Their bodies burned as I watched. The fire must have burned for more than an hour. I still cannot forget that sight!"

The Long Battle Begins

This was the beginning of Margaux's long battle with both TBI and PTSD. Her face developed Bell's palsy, a paralysis or weakness of the muscles on one side of the face, which she now thinks came from the mental stress she had suffered. As she recalls, "The doctors thought maybe this was happening because I was in the third trimester of a pregnancy or because I had slept on my face wrong … or due to some other cause. They were idiots and just wouldn't accept that it was due to stress."

Eventually she was sent back to Landstuhl. In April 2007, she began receiving further medical attention and the doctors finally realized that her initial concussion and mental state were more severe than originally diagnosed. Recalling those dreary days, Margaux describes her feelings, "I had atypical nerve pain and trigeminal neuralgia. At that time my post-traumatic stress was off the charts. I realized I would not be returning to Iraq and was feeling like I had abandoned my team. I felt like I was a traitor. When I left them, I told them I would be back in a couple of weeks." Her unit was deployed for a full thirteen months and she would not see them until they returned to Germany. She vowed to them through email that she would not drink or celebrate anything without them. "I did not touch alcohol until they returned. I was in so much pain anyway that I couldn't drink if I wanted to, so I slept fourteen hours a day every day in Bamberg."

Margaux spent her days visiting the hospital three hours away from Bamberg and there she received a multitude of pills—morphine, Percocet, and more medications. Then the doctors decided her PTSD was so bad that they sent her home to Colorado because that was where her family lived.

Remembering that journey home to Colorado and checking in, Margaux says the Army doctors gave her a choice, "Stay on pills for the rest of my life or try brain surgery." She finally had brain surgery in March 2008. That was the beginning of the end of her active Army duty.

"The surgery was performed to try and alleviate the pain. I was in so much pain I couldn't even walk up a flight of stairs. It felt like a softball had been inserted inside of my skull. And so I was willing to try anything … that is when I underwent laser surgery at Fort Carson, Colorado, where I would remain for eight months. But it was unsuccessful. The severe headaches continued, and I was medically discharged. Then in 2009, I planned to move to Ohio with my fiancé, but before doing that I had the great fortune to meet a Fort Carson Army representative who assisted wounded warriors like me through the Warriors Transition Unit. Robert B. Alvarez, a Marine Corps veteran himself, interviewed and processed many wounded warriors like me. He told me he would help me get my life back. I had suffered hard from both the brain injury and the post-traumatic stress, and it was just fate that I ended up in his office. If I had never met him I would never have gone through hyperbaric oxygen therapy. It was then that Bob Alvarez introduced me to Dr. Paul Harch, the pioneer of HBOT therapy who had his clinic at Southern Louisiana University."

Margaux in the hospital
~ courtesy of Margaux Mange, U. S. Army (Ret)

A Lucky Break

In 2009, Margaux became one of Dr. Harch's patients and traveled there to receive HBOT treatment. Margaux tells of the experience, "There I took forty *dives*, and I have to say it was successful. I started feeling better after about twenty of the sessions. However, I was angry. At the same time I was getting treatment, my Ohio psychiatrist had also sent me there with home-work. I have always been a faithful-like patient and listened to what a doctor would tell me. I have always tried to do the right thing and didn't like people being mad at me. Then, during the first part of the treatments my anger surfaced. I looked at the homework and it made me so angry. I was so ticked off at her that I ripped up everything I'd brought. All of those emotions started coming out that I had been hiding. My memories were now resurfacing and I was crying and laughing a lot more."

Dr. Harch had warned Margaux that after a few sessions of hyperbaric chamber dives, while her brain was healing physically, her mind would also experience a rollercoaster of emotions during the final recovery process. Remembering that, she recalls, "When I came back and talked with my psychiatrist I handed her my torn pages of homework. She looked at me and said, 'This isn't you.' With all those terrible emotions coming to the surface we were actually able to figure out together what was going on with my post-traumatic stress situation." When Margaux returned after that session with her psychiatrist, she describes how they began "digging into new stuff and it was really helping me a lot. Before then, I hated all of my emotions. I was a robot and numb before the HBOT treatment with Dr. Harch. I didn't like that I had to go through the treatments all by myself. But after my first forty treatments, I was able to deal with the terrible memories of what had happened to my best friend, Ashley, when she died on March 3, 2007."

In February 2010, after Margaux completed her forty dives with Dr. Harch, they determined that her TBI and PTSD were more serious than the mild TBI first diagnosed. This would require many more chamber hours, which Robert Alvarez found in Louisville, Colorado, close to my father. In August 2010, Ryan and Eddie learned of her amazing story and offered to treat her at the Rocky Mountain Hyperbaric Institute "for as long as it takes to heal you." So, another forty chamber hours were scheduled over the next few months. Margaux continued to return for additional "maintenance" treatments for several years after that (achieving a total of 140 HBOT chamber hours between both facilities). Today Margaux knows she can finally cope with her TBI and its related PTSD. She still volunteers to raise funds for the Rocky Mountain clinic.

Breakthrough

Margaux tells others that the only negative aspect to this treatment is that a person must lie in a tube for an hour every day and that she thinks that people are just impatient with the need to do that. She also thinks many people still don't know about this form of treatment and its benefits. When people are new to the treatment, there are still those who find it hard to believe in its benefits. Taking a pill provides quick relief, but the effects are temporary. Sometimes it is difficult to get people to understand that pills won't work in the long run, but the long-term effects of hyperbaric oxygen treatment will be well worth the time it takes.

Margaux discovered that the healing began to happen after about fifteen to twenty sessions. At that point, her ability to sleep improved, her migraines lessened, her memory began to return, and her depression and need to isolate and withdraw from others eased up. Her physical wounds improved with HBOT, and the TBI started to heal. The post-traumatic stress, however, was an awful situation for her, and as her memory and brain functions began to improve, she had a bad PTSD slip, a brutal one, as Margaux describes it. "You go into a deep dark depression where all of your emotions go down. If you are an avoider, you avoid everything, but you can't avoid the PTSD forever. All of my emotions, every single one of them came to a boiling point. I don't know how to describe it, but your mind is not strong enough to shut down all the negative emotions you experienced."

Even though she was physically healing her brain from the concussion, she needed professional help with the mental and psychological wounds she had also suffered. "When I was first told by the doctor in Germany that I had post-traumatic stress disorder and I would not be going back to Iraq or any combat duty, it was the lowest point in my life. You know, I did not know what it was, this PTSD they were talking about. It was devastating for me ... I had been planning on a career with the Army."

Now that Margaux has gone through over 140 hours of dives, she is not the robot that she thought she might become. In fact, her own inner strength and competitive nature have brought her back to the Margaux she always was.

Despite dealing with the signature wounds she suffered, she now excels in mountain climbing. She has scaled 19,000-foot Cototaxi in Equador and a more recent ascent of Mount Denali in Alaska that was thwarted by weather at 16,500 feet. Marguaux and other veterans returned to the summit (20,300') in 2016 reaching it carrying four American flags inscribed with 500 names of veterans ranging from WWII to Afghanistan. She participated in the cyclists' Ride to Recovery that toured Washington State and the coast of California as well as rode bike tours through Italy and France.

Since 2011, Margaux has competed three times in the Wounded Warrior games held at Colorado Springs. She won five gold medals and three silver medals, just one more proof that our former athlete still has it.

To top it off, Margaux was chosen to represent America's Walking with the Wounded team that competed with a European team headed by Prince Harry of Great Britain and an Aussie-New Zealand team in the South Pole Allied Challenge that began November 14, 2013. Not only did Margaux and another American female warrior race against the men of the other teams, but one of Margaux's team was a blind warrior who needed assistance every step of that long and difficult Antarctic trek.

Once again, the true Margaux was revealed when an opportunity arose and she was able to guide the blind warrior during several parts of the trek. She had asked to do it because she knew she just had to. It was her own small contribution to give back what she had been given. She so vividly relives those minutes and hours. "I asked if I could guide Ivan, a blinded warrior. I got the privilege of guiding him and I didn't want any cameras or anyone around … no pictures. I did not want anything to be about me. I wanted to understand how he had been suffering and I wanted to be able to help. And I enjoyed every minute of helping him! I was working up a sweat, because I was exerting myself beyond normal, but it felt so good to work and struggle and do this for Ivan."

Margaux met the prince and all the international participants in England, where they were wined and dined before flying out to New Zealand. They then sailed off to the race site in Antarctica. Their arctic trek was filmed from the beginning, and it was obvious that Prince Harry was out to win it all for his team. Cameramen accompanying the racers had as much difficulty with the rugged, inhospitable terrain as the racers obviously did. Within the first few days, Harry had managed a considerable lead over the other two teams. It became apparent, however, that something was very wrong. Filming revealed that the American team was falling behind by quite a distance. When Prince Harry realized it was due to the team's challenges to assist their blind team member, he decided that all three teams would race together and Margaux and her friends would get help with their handicapped racer.

Everyone at the Rocky Mountain clinic admires Margaux for her exceptional humility and spirit. Eddie and Ryan were extremely proud when she displayed the Rocky Mountain Hyperbaric Institute and Rocky Mountain Hyperbaric Association for Brain Injuries banners at the pole. Margaux's performance and gutty grit in that grueling race set a new level of fortitude and courage that this lovely, young combat veteran consistently demonstrates in her life.

Margaux with Prince Harry in Trafalgar Square
~ courtesy of Margaux Mange, U. S. Army (Ret)

The Rocky Mountain Hyperbaric Institute also owes a major debt of gratitude to Margaux for being the first to step up and share her personal experience with other veterans who seek treatment. Very large donor groups have heard her compelling story, and today it is their generosity that provides the funds to treat hundreds of combat veterans free of charge. She knows they too can be healed in the same way she was. Margaux is quick to credit Robert Alvarez, counselor/therapist, for coming into her life and helping her with her new life. Today Margaux attends the Colorado Mountain College in Leadville, Colorado, ready to climb even higher.

Margaux arrives at the South Pole with
Rocky Mountain Hyperbaric logo
~ courtesy of Margaux Mange, U. S. Army (Ret)

DOD Photo (USMC) 151007-M-SV584-059
~ courtesy of Corporal Anna Albrecht

In 2015, a very unusual request came to the Rocky Mountain Hyperbaric Institute. Our good friend and supporter, Rob Beckman, called Eddie to ask if we could schedule an active duty Marine into the HBOT program, quietly and anonymously. It seemed that Rob's son had been approached by a Stanford classmate who knew the wife of a Marine Corps Gunnery Sergeant who was suffering serious TBI and PTSD. Her tearful request to Rob saved this Marine's career at a time when the Corps would have discharged him for his inability to perform his regular duties.

This was usually the situation with Soldiers, Sailors, and Marines who suffered the signature wounds of war, but his compassionate Marine commander, a mustang officer who knew how seriously his Gunny was suffering, had covered for him right up to his re-assignment to another duty station. There, his commanding officer knew that the rigors of this new job as an instructor would result in his obvious inability to perform his duties satisfactorily. Once his condition was discovered, the command would have no alternative but to medically discharge him. With thirty days' leave on the books, his commanding officer first tried to have him treated at a nearby Navy hospital that had HBOT chambers. Unfortunately, those doctors told him they were constrained by the surgeon general from treating any TBI with HBOT.

Rob knew about the Rocky Mountain clinic and its success treating TBI and PTSD. He advised the CO that Ryan and Eddie would provide the Sergeant's treatment free of charge. All he had to do was come to the clinic in Colorado, where he would be treated during his leave period and boarded and fed at their nearby veteran's home, also free. Everyone at the clinic believed the Gunny's career could be saved. After his HBOT treatment, he would be well enough to carry out his next set of orders.

In order to get started, he had to contact the clinic's medical director, Dr. Julie Stapleton, to be interviewed, screened, and have his prescription written. Eddie advised the Gunny that the clinic could process him in the regular forty chamber hours, but that they would double up the hours in the last ten days with two treatments per day. This shorter treatment time span had worked for other veterans, and the clinic believed it would also work well for him.

Force Recon Marines in action
~ courtesy of DoD (USMC) A338299, photo by R. E. Endris

In every military service, a special type of warrior is trained for combat that is unique and very difficult. Only a small percent become Navy Seals, Special Forces, Air Force Commandos, and in the Marine Corps, Force Reconnaissance Marines. Their small unit operations are designed to be dangerous and at close quarters where close-in firing and mortar-grenade blasts are common. The fact that they experience more severe TBI and PTSD comes with the job and the deadly missions they carry out secretly. Gunny "T" was one of these special Marines.

Gunny "T" had completed seven combat tours in Iraq and Afghanistan and survived seven major concussive blasts. But, as his worried wife explained to Rob, he now suffered severe head pain and lengthy migraines, memory loss, little or no sleep, and was deeply depressed over the suicides of five recon teammates who had suffered the same physical and mental problems he did. She was worried that he might become the sixth. The Gunny wrote his own report for Eddie, who began conducting cognitive and behavioral testing after the first HBOT treatments:

"My Force Recon unit has literally been *hollowed out*, with so many of us carried on the official roster but non-deployable because of our serious mental and physical conditions. Many of us have TBI and PTSD. I have served in the military for fifteen years and have received several head injuries. The first two I received in Fallujah during 2004 and 2005. This was before the military even began to conduct studies on head trauma or battlefield concussions.

"In one vehicle patrol, I hit a roadside IED and was knocked unconscious. The next one happened just two days later on another patrol, while I was still suffering the symptoms of the first IED. During the next thirteen years, I received three more major injuries from breaching operations while conducting special missions in Iraq and we were danger close to ordnance [bombs] dropping very near us. My last severe concussion was during regular Recon training, when I had a Jet Ski boating accident. That is when my symptoms got to such a point that it was affecting my personal life and my marriage.

"I was in and out of the emergency room and doctors' offices for four years and they all said and did the same thing: 'Headaches are really hard to diagnose, so here, try these pills.' At one point I was given six different medicines and I felt like I was in danger of losing my job in the Corps. In order to do my job and try to function normally, I had to stop taking the meds. So I dealt with the painful migraines three to six times a day. At my worst point, I would suffer a very painful migraine for as long as six days, then have a short break with a less severe headache, and then the bad migraines would come back. That is when I became seriously suicidal because that pain drove me there.

"Thanks to HBOT treatment at the Rocky Mountain clinic, I have been without any of the symptoms for four weeks now and my ability to multi-task has returned. My wife calls my recovery a miracle. My cognitive ability and thought processes have increased considerably, from about 35%, my own self-measure, to 85% today. And my mood and performance of military duties is 100%. I am no longer viewed as *The Anger Gunny*."

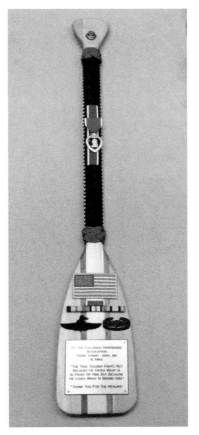

Presentation of the "Force Recon Paddle" inspired by Gunny "T" and three healed warriors for Rocky Mountain Hyperbaric Association for Brain Injuries
~ courtesy of Grady Birdsong, Veteran Advocate

Staff Sergeant Ian Newland, U. S. Army (Retired)

Sergeant Ian Newland in Iraq with 1st ID, 1/26th Inf. Bn.
~ courtesy of Ian Newland

Imagine riding in your Humvee troop carrier in one of the deadliest sectors of Baghdad, on December 4, 2006. You and four other soldiers were assigned a difficult and dangerous mission to restrict enemy movement and quell sectarian violence in Adhamiyah. As members of the 1st Platoon, C Company, 1st Battalion, 26th Infantry Regiment, your job was to protect a convoy's rear as it maneuvered through the narrow streets.

Pfc. Ross McGinnis was manning the turret-mounted M2, .50 caliber machine gun, when an insurgent positioned on a nearby rooftop tossed a fragmentation grenade into the open roof of the vehicle. Without hesitation or regard for his life, Pfc. McGinnis quickly threw his back over the grenade and absorbed the lethal and concussive effects with his own body.

Riding in that same vehicle was his team leader, Sergeant Ian Newland, who also received many serious wounds from that exploding grenade. A large piece of its shrapnel hit him in the head and knocked him unconscious. But he soon revived and tried desperately to lift Ross's body off him. It was then that he saw the severity of the grenade wound and knew it had killed his brave friend. All of the soldiers within the vehicle were injured and were medically evacuated to the Army hospital, where doctors then determined that Ian's body had received more than fifty pieces of grenade fragments. These were his career-ending wounds and the beginning of years of hospital treatment and medications.

Born in Dayton, Ohio, where he was raised on a farm, Ian was deeply moved by the 9-11 terror attack in New York and decided to join the Army in 2002. After basic infantry training at Fort Benning, Georgia, he was assigned to the 1st Infantry Division, 1-26 Infantry Battalion in Schweinfurt, Germany, his home base, until March 2004, when his "Blue Spaders" deployed to Ballad, Iraq, in support of OIF-II.

During this deployment, Ian earned three Army commendation medals, which included one for valor in the battle of Samara. After returning to Germany as a Specialist, he re-enlisted as a career Soldier with the intent to be a "lifer." He completed Air Assault, Airborne, Warrior Leader's Course, Pre-Ranger, NCO PLDC, and sniper training. In 2005, he was promoted to Sergeant and selected to compete in the 1st Infantry Division Non-Commissioned Officer of the Year competition.

In December 2006, he was re-deployed with the 1-26th back to Iraq. During this second deployment, Sergeant Newland was promoted to the rank of Staff Sergeant and received a Bronze Star medal for Valor, two Bronze Stars, a Meritorious Service Medal, an Army Achievement Ribbon, and the Purple Heart.

His unit was the hardest hit Army unit in Iraq in twenty years. They suffered seventeen casualties in Ian's company alone. That combat account is described in Kelly Kennedy's book, *They Fought for Each Other: The Triumph and Tragedy of the Hardest Hit Unit in Iraq*. But that tragedy would become even worse for Ian as he continued fighting the brutal, bloody war and then lost his best friend to an enemy grenade.

In subsequent interviews Ian spoke about his buddy Ross McGinnis. "We had to replace two key NCOs who we lost, but even as a PFC it was soon obvious he not only carried his own weight but immediately became an instrumental part of our team. His special personality and humor made him stand out. He was the comedian of the platoon and he made your worst day livable and had us all laughing within minutes. Our Platoon Sergeant had assigned him as our .50 caliber gunner because he was such an outstanding soldier in every way. I still have this image of him, the funny guy with the infectious smile, often smiling down from his gunner position, and his teeth caked with the mud that spattered him. But that smile made our day.

"December 4th was just another day and we were delivering a generator to provide electricity to that area. We knew insurgents had been lobbing grenades at our vehicles, but our reaction drills let us ward them off. That day, Ross was looking out the rear of the Humvee and covering the rear of our convoy with his machine gun, when the insurgent managed to toss the grenade right at him. It bounced off his helmet and landed on an ammo can below him. Ross knew we could not escape from the vehicle, which was under enemy fire, and instinctively he rolled back over the grenade to absorb its blast with his whole body. I tried to lift him up, but I could see that the huge explosion had killed him. He died saving the rest of his pals. That is the kind of soldier Ross was. And I am alive today thanks to his bravery. My wife still has a husband and my kids still have a father. There is not a day that goes by that I do not think about his courage."

★ ★ ★ ★ ★

PFC Ross Andrew McGinnis, Medal of Honor recipient

Ross McGinnis initially received the Silver Star for his individual heroism and subsequently was awarded the Medal of Honor by President George W. Bush at a White House ceremony that Staff Sergeant Newland attended. Burial at Arlington Cemetery followed, when Ross was given a hero's honors and burial ceremony. The loss of his close friend was overwhelming for Ian, who had not only suffered severe disabling physical wounds himself, but lived with the sad memories of this remarkable, fun-loving soldier. Those memories haunted him so much that he soon experienced the serious grief and guilt associated with PTSD. This combination of signature wounds would be his cross to bear for years, until he learned that there was a healing program for combat veterans.

Ian became one of the first veterans to be treated and healed under Ryan and Eddie's "Healing Our Heroes" nonprofit program and 501c3 fund. In an exit interview, Ian openly and frankly stated that his HBOT therapy saved his life, and it was almost a miracle that he had his health and life back. Gone were his crutches, his cane, and so was his service dog, who had been trained to nudge and lift him back on his feet after he fell during those terrible days when he still suffered with many physical wounds. So impressed was he with his HBOT treatment that he volunteered to participate in the early video Grady Birdsong produced about the clinic and its veterans.

That video impressed many donors and, as a result, the clinic was able to raise enough money to move to its current location in Louisville, Colorado. What had been basic, everyday survival became a more livable, enjoyable life as Ian returned to his family and his friends. He was eager to continue his new life as a civilian. Ian was retired with the rank of Staff Sergeant with four years of service, and he left the Army with a medical discharge.

Ian Newland in HBOT chamber
~ courtesy of Grady Birdsong, Veteran Advocate

Staff Sergeant Newland spent five years rehabilitating his severe wounds and completed his HBOT therapy in 2011. Never surrendering to his injuries or to the sad loss of his close friend, he became an avid public speaker and supporter of many wounded warrior and other veterans' programs all across the United States. He also began training to become an amateur athlete and completed the Army 10 Miler, World T.E.A.M. Sports Adventure Team Challenge Colorado 2014, and the 30 kilometer "Ruck Race."

He also served as president of the Honor and Courage Mountaineering team and describes, "I look forward to the World TEAM Sports – Team Adventure Challenge every year. It still challenges me mentally, physically, and spiritually. The most powerful aspect, though, is that it's not just about me. When I was injured, I came home and lost the comradery and the teamwork, along with the feeling that I had my buddy's back and he had mine. When you come home from war and you're separated from service, it can be lonely. The Adventure Team Challenge provides veterans with the opportunity to feel like we belong to something bigger than ourselves again, to gather a team and work towards a common mission, to find collective success, and, most importantly, to fill a crucial role where you know you need others, and others need you."

Staff Sergeant Ian Newland, U. S. Army, Retired
~ courtesy of Ian Newland

Major Ben Richards, U. S. Army (Retired)

It is a rare privilege to meet a combat veteran of the Iraq War who not only displayed exceptional leadership in his worst days on the battlefield but whose remarkable recovery from his own signature wounds is why we include his story here. Ben has volunteered his time to raise funds for HBOT treatment of veterans suffering TBI and PTSD. He joined General John Allen USMC (Ret) at the April 1, 2015 Fundraising Gala in Denver, Colorado, and his own personal account left few with dry eyes.

Today he is a spokesman for Soldiers for the Truth (SFTT), an organization founded by Vietnam veteran Colonel David Hackworth and now chaired by his wife, Eilhys. It was SFTT who learned about Ben's disabilities and proposed his treatment by Dr. Paul Harch. Ben received forty hours of HBOT therapy at Dr. Harch's Health Sciences Center at Louisiana State University in 2012. Today Ben joins the hundreds of veterans we have treated at the Rocky Mountain clinic. It is our pleasure to tell his story.

From an early age, Ben wanted to go to West Point and become a soldier. He was a top student, exceptionally bright with an IQ of 148. Becoming fluent in Chinese, he graduated in 2000 as a 2nd Lieutenant and looked forward to a long Army career. Ten years later, Ben returned to become an instructor in the history department at West Point, but not before experiencing the worst days of his military life. He found himself in a place where the enemy fought a different kind of war.

In 2006, Captain Ben Richards, whose young Army career was not only stellar but had been cited by his seniors as one of the finest officers they had ever commanded, deployed to Iraq. In 2007, he took command of a Stryker cavalry troop consisting of 100 soldiers and 17 armored vehicles. His "Bronco Troop" was assigned to patrol and control the hot-bed city, Baquba, where he was immediately welcomed with a twelve-hour firefight the first day and quickly earned his Combat Infantry Badge.

There he established a controversial cooperative program with Sunni Muslim militias who had previously attacked American units. Subsequently, they helped Ben's unit defeat a worse enemy, Al Qaeda. Even though he was initially criticized, his strategy worked and was adopted in the U.S. military all over Iraq. The Army promoted him to Major.

It became routine for his Stryker team to encounter numerous IEDs and deadly blasts as well as suicide car bombers. His unit's actual experience was filmed on YouTube and shows the deadly, destructive IEDs exploding as they advanced. In that spring and summer, his troops had excellent officers and NCOs—five of his officers were also West Pointers.

Major Ben Richards, USA (Ret)
~ courtesy of Rocky Mountain Hyperbaric Association

Ben tells of their first missions, "During those few months, 90% of my men hit at least one IED—often more than once. That May 2007, a suicide bomber driving a sedan laden with explosives rammed into my Stryker and caused so much damage to it that it was beyond repair. For most of my troops they were earning their Combat Infantry Badge almost on their first day."

One of the deadly devices blew up under his vehicle and gave Ben a severe concussion that left him dazed and sick for a week. Three weeks later, another roadside bomb blasted his vehicle and knocked him out. Another concussion left him in even worse condition, with piercing headaches, a TBI, and post-traumatic stress. He felt that he was not really that badly injured since he still had all his limbs and had not suffered wounds like World War II and Korean veterans did when they received Purple Hearts. He returned to Fort Lewis, Washington, and six months later was diagnosed with PTSD. He only sought help when he was *command directed* by his wife, who recognized his condition.

It was Ben's wife, Farrah, who first noticed the major change in her once vibrant husband when he returned to regular peacetime duty. He seemed withdrawn and totally aloof from his former activities and the roughhousing he used to enjoy with his kids. He no longer played with them and he certainly was not the active, alert, and interesting man she married. While sleeping he seemed to be on a hair-trigger and would be startled awake at the least noise. She feared he might actually see her as the enemy and harm her in some way.

Their marriage became strained, and they both felt they might have to separate. There were obvious changes in his personality as he suffered migraines, slept little, was unable to concentrate or speak, and could hardly write the same Chinese he once knew so well.

Even when he was sent to Georgetown University for his master's degree, his professor, Nancy Bernkoff Tucker, commented on one of his term papers about China, "Parts of it are not coherent and overall it is not effective. It is not well written but quite sloppy. But of greater significance is its lack of analysis." Ben had to agree with her. He knew his reasoning capacity was impaired and his memory and ability to concentrate were faulty. A brilliant man was failing. It took three years after his Iraq tour for Army doctors to diagnose that he did have TBI and PTSD.

The symptoms of his signature wounds were beginning to show, but back then there was no effective treatment. Struggling daily from these wounds, Ben was then assigned to the history department at West Point as an instructor in the summer of 2010. He soon realized, however, that he could not measure up to that challenging job. He lacked the dedication and desire to teach, and he knew there was something very wrong with his mind and his memory, so he asked to be relieved from his teaching duties. While his history department bosses covered for him and realized that he had suffered unusual brain trauma and PTSD, his fellow instructors, both civilian and military, also took on much of his workload, trying their best to help get him the care he obviously needed.

After a battery of physical and psychological tests and a spectrographic scan found eight lesions in his brain, the Army doctors determined that he was disabled. Ben was forced to retire after thirteen years of service. As he recalls that terrible day, "Their best care and concern was not enough and they had tried everything to ensure my future well-being. I received their farewell honors and best wishes. I knew my career had ended, but not as I had ever planned."

He left the Army a virtual basket case and subsequently encountered the futility of the VA, whose care and concern for veterans with these wounds was inadequate. Ben was fortunate that other veterans stepped up to help, like his Army friend and member of the Stand for the Troops (SFTT), who arranged HBOT treatment with Dr. Harch at his Louisiana clinic.

Dr. Harch and his wife, who was also his business partner, paid for his forty-hour treatment program, and the SFTT paid his way to the clinic. They also paid Ben's room and board during his treatment period. He had heard about HBOT and what it was doing for other veterans at the time. Ben was more than willing to take on any beneficial treatment that would eliminate the endless pills prescribed.

Today Ben credits Dr. Paul Harch for his remarkable recovery and has been an avid supporter of HBOT. He has visited the Rocky Mountain Hyperbaric Institute and received "maintenance" treatment in their chambers. He even volunteered to help raise funds at the "Healing Our Heroes" nonprofit fundraiser on April 1, 2015, at the Hyatt Regency Hotel in Denver, Colorado. Joining Retired General John R. Allen, United States Marine Corps, that evening, Ben gave a passionate and moving account of his own wounds and HBOT treatment. Thanks to Ben, Margaux, and General Allen, the program raised enough money from the 250 attendees to fund veteran HBOT treatment until 2017.

Ben's contribution does not end there. He recently promoted a Colorado Springs nonprofit company called the Brain Health Team and its Low Level Light Therapy (LLLT) that compliments HBOT in its healing of TBI. He also introduced Dr. David Dubin and his Low Energy Neuron Stimulator (LENS) that has recently been successful in treating the brain trauma and concussions suffered by retired NFL football players like Joe Namath and others.

Just as other doctors and psychiatrists are searching for new healing methods and techniques, Ben and this team are now opening doors to these important therapies that he himself admits using. As he recently said upon leaving the Rocky Mountain clinic, "I have not had a single sign or symptom of PTSD since I used both of these devices." Yes, it is a miracle!

CHAPTER 9

Patient Treatment Program

Ryan Fullmer describes the evolution of the clinic's treatment plan, as follows: "When Eddie and I set up our HBOT clinic and nonprofit fund to treat and heal combat veterans more than five years ago, we began without a comprehensive treatment plan. We had no measures of successful treatment other than the veteran's personal testimonies. Today, we have in place a treatment plan that includes both qualitative and quantitative measures to support the successes of our HBOT program. That same plan applies to our civilian patients."

Rocky Mountain Hyperbaric Institute Treatment Plan

1. Any individual who is suffering from concussion, stroke, or brain damage of any kind may apply for treatment by filling out our Application for Treatment Form. (Form available at www.rmhabi.org)

2. A veteran's application must include the DD-214 or Discharge Certificate, which will be reviewed by the RMHA Board of Directors. Once military service records are verified, the applicant is notified of approval or rejection.

3. Once approved, the patient on arrival will undergo a medical examination and screening by the RMHA Medical Director, Dr. Julie Stapleton, to ensure they have the fitness and general health to undergo compressed oxygen therapy at the clinic. Once approved by Dr. Stapleton, she writes the prescription to receive hyperbaric oxygen treatment.

4. Prior to actual chamber treatment, a series of tests are administered to establish a physical and emotional baseline of each patient. They include quality of life, post-traumatic stress evaluation, and other neurocognitive behavioral indicators. Depending upon the severity of their TBI-PTSD, these tests are administered at key times throughout the treatment program.

5. For traumatic brain injured veterans, CereScan Corporation of Denver, Colorado, offers periodic Brain Scan Imaging (before and after the forty-hour treatment period) that provides horizontal, sagittal (divides the brain into left and right), and front and back views that clearly show the efficacy of hyperbaric oxygen infusion. The results of a 3D SPECT scan are displayed as shown in the example below. Brain SPECT imaging is a sensitive measure of cerebral blood flow and evaluates functionality in up to 160 regions of the brain (See Exhibit D to learn more).

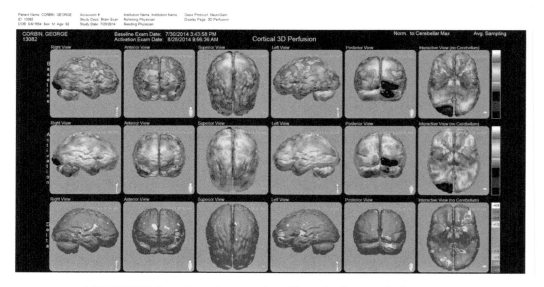

A 3D SPECT Scan imaging result utilized before and after HBOT
~ courtesy CereScan & Colonel George Corbin, USA (Ret) http://www.cerescan.com

6. The veteran meets with Pepe Ramirez, PTSD Program Counselor, who evaluates each veteran for their specific degree of combat trauma

and then initiates his integrated cognitive and physical therapy program that parallels the HBOT chamber treatment. PTSD status, including regression, is continually monitored and reported.

7. The overall treatment program consists of forty hyperbaric oxygen chamber hours and twelve PTSD counseling hours. Each treatment period is approximately one-and-a-half hours in the chamber that include approximately fifteen minutes to compress and then fifteen minutes to decompress.

8. After the fortieth treatment hour, the patient retakes the original series of quality of life evaluations and the neurocognitive tests. The results are then evaluated and the Board of Directors will determine if a patient is eligible for more treatment.

9. PTSD Counselor Ramirez conducts follow-up evaluation of each veteran to determine if he or she needs further PTSD counseling and therapy at their local VA Clinic or Hospital.

Rocky Mountain Hyperbaric Institute Treatment Sequence

In describing the hyperbaric treatment sequence, Ryan tells all newcomers that after successfully treating more than 250 veterans and hundreds of civilian patients as well, "Eddie and I are confident that we now provide a comprehensive, integrated and very safe hyperbaric oxygen therapy program, which is tested and measured for positive results. Our treatment sequence demonstrates this.

"After the application for HBOT therapy and verification of the patient is completed, we schedule them for a standard forty-hour program, as is depicted in the following sequence."

Dr. Julie Stapleton interviews and examines the veteran to determine if there are medical reasons to reject him or her for HBOT therapy, such as bronchial illness, lung disease, or COPD. Once approved, she writes the prescription for hyperbaric treatment. Using her three-step approach, she first makes sure the patient can undergo the chamber pressure. Second, she answers questions about the process, reassuring the individual that the entire process is safe. Third, she educates the individual on how to clear their ears while under pressure. Most of the veterans Dr. Stapleton has examined to date have successfully undergone the complete treatment process.

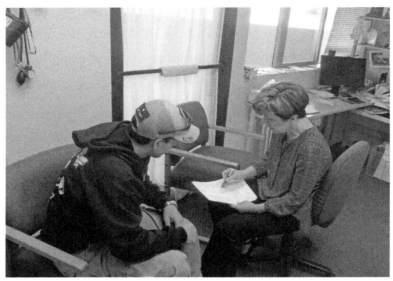

Dr. Stapleton prescribes treatment
~ courtesy of Grady Birdsong, Veteran Advocate

Eddie Gomez (right) reviews process with Veteran
~ courtesy of Ryan Fullmer

Eddie Gomez introduces the veteran to the HBOT program and describes the oxygen chamber procedure, dive time, compression-decompression, the oxygen treatment process, and the safety procedures that will be strictly followed. It is at this point that the veterans take tests to identify cognitive deficits that have resulted from concussions or traumatic injuries. The initial data captured will be compared and contrasted when the veteran retakes these same tests post-treatment.

Three-bedroom home for visiting Veterans
~ courtesy of Ryan Fullmer

Some veterans who travel to the clinic from out-of-area locations reside at the house located nearby and provided by the clinic's nonprofit fund. Most of those patients engage in two sessions a day—one in the morning and another in the afternoon. When multiple veterans are visiting, their schedules are co-ordinated in advance. The house is a short two blocks walking distance from the clinic.

One of the Rocky Mountain Hyperbaric clinic assistants positions the oxygen hood, which will concentrate and infuse oxygen into the damaged areas of the brain to stimulate new blood cell and blood vessel growth (angiogenesis) during the one-hour session. Hospital scrubs (sanitary clothing) are worn by all patients during every chamber session. These garments are designed to reduce contamination and are laundered after each session.

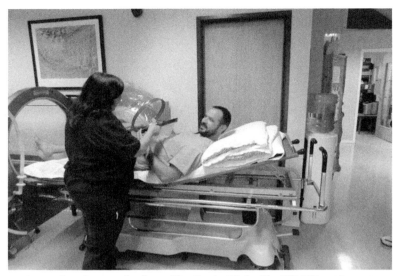

Introduction to the hood by Josie Nicholls, HBOT assistant
~ courtesy of Ryan Fullmer

Patty Juarez, HBOT assistant, preparing for the "dive"
~ courtesy of Ryan Fullmer

An HBOT assistant then prepares the chamber for compression. The "dive" is administered at the pressure of 1.5 ATM, which is equal to 17 feet below sea level. This depth of pressure is appropriate for the "Mile High" (5,280 feet) density altitude in the Rocky Mountains of Colorado. Oxygen breathed under pressure forces more oxygen into the damaged tissue area of the brain and encourages new blood vessel growth, as well as triggering a cascade of gene expression geared toward healing and promoting brain repair. This process of compression is performed each time a patient enters the chamber. It takes approximately twelve to fifteen minutes for the patient to reach the desired chamber pressure of 1.5 ATM. The patient then puts on the oxygen hood once chamber pressure has been attained. The patient remains chambered for one hour. The process of decompressing after one hour in the chamber also takes another twelve to fifteen minutes. One session requires about two hours, from the time the patient arrives and enters the chamber until the dive is completed and they exit the clinic.

A "dive" in the chamber
~ courtesy of Grady Birdsong, Veteran Advocate

This "sit-in" chamber infuses the veteran with 100% oxygen while he or she reads a book, watches a video or movie of their choosing, or just rests during the one-and-a-half to two hour chamber time. A technician attends to each patient and is always in contact by intercom and conducts visual checks throughout the time the chamber is under compression. After spending an

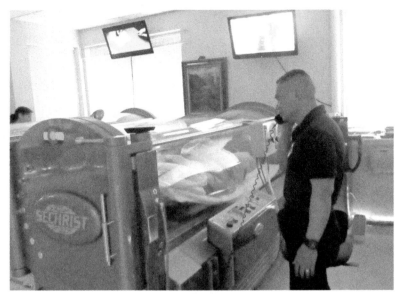

Counselor Ramirez checks Veteran status
~ courtesy of Ryan Fullmer

Ever vigilant, Trauma counselor Pepe Ramirez regularly monitors each veteran as they progress through the entire HBOT treatment process, watching for any signs of regression (any negative PTSD reaction). Depending upon what he observes, Ramirez may schedule additional cognitive or physical therapy in future sessions. The patient can terminate a chamber session at any time he or she desires to do so.

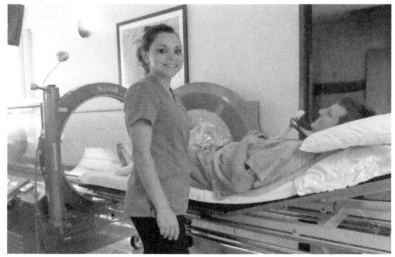

HBOT assistant Patty Juarez assisting Veteran after decompression
~ courtesy of Ryan Fullmer

An HBOT assistant assists the veterans after each decompression from the chamber. She removes the hood and lets him or her adjust to the altitude of the room. She will alert a certified hyperbaric specialist (CHS) should the patient exhibit any signs of distress.

Post-assessment evaluation
~ courtesy of Ryan Fullmer

"Paper and pencil" tests, which are administered pre- and post-treatment, identify cognitive deficiencies and levels of improvement. The CNS-Vital Signs computerized neuropsychological screening exam is also administered to assess brain functioning, such as short-term memory, attention, concentration, visual spatial capacity, processing speed, and reaction times. This has become an important pre-treatment and post-treatment tool that provides a reliable metric for describing effectiveness of the treatment. All tests are used in assessing severity and outcomes. (See Exhibit C on the CNS-Vital Signs clinical testing software.)

PTSD Debriefing

Veteran Cody Casteel during his debriefing session
~ courtesy of Ryan Fullmer

During debriefing sessions, the veteran discusses his mental and physical condition as it relates to his PTSD therapy and progress during his stay at the clinic. From these interviews, Counselor Ramirez determines what additional PTSD counseling and therapy a veteran may need at a local VA hospital or clinic when they return to their home. By the end of a veteran's treatment at the Rocky Mountain clinic, Counselor Ramirez has coordinated with the veteran and his family to ensure that resources and a plan is in place when the veteran leaves, in order that the veteran can continue to receive the same level of cognitive therapy treatment in his or her "hometown." Pepe Ramirez works to make that transition as seamless as possible.

Regression can occur with a patient when HBOT begins to physically heal the brain. This is common among combat veterans who suffer from both TBI and PTSD. The negative elements of PTSD (grief-guilt-anger-withdrawal-depression) are often suppressed but can re-surface as the brain begins to heal. Recognizing this during the early days of the clinic's HBOT program has resulted in Eddie and Ryan's integration of the PTSD counseling program into their existing hyperbaric therapy. It has made a world of difference in outcomes!

As the Rocky Mountain Association approached its 300th veteran, Ryan made the decision to expand the clinic by adding treatment rooms for alternative therapies and more office space. In part, some of these new therapies are addressed and discussed in this book by Dr. Julie Stapleton and Pepe Ramirez. Dr. Xavier Figueroa expands on them in his chapter, *Afterward: The Path Ahead* beginning on page 147. Major Ben Richards, a TBI patient additionally benefited from Low Energy Neuron Simulator (LENS) treatments which have been successful in addressing concussion in the NFL. Just as the clinic's innovative integration of PTSD counseling and physical therapy have proven successful, now the inclusion of alternative treatments will be studied and employed by the clinic and its unique miracle working team.

CHAPTER 10

Clinic Measurement Tools and Patient Progress

In conjunction with this book and its authors, Dan Guenther, author, veteran advocate, Vietnam combat veteran, and former Director of Research for International Learning Systems, Inc., conducted a professional review of the Rocky Mountain Hyperbaric Institute's processes and its measurement methodology. The purpose of this study was to determine and document how the measurement tools are utilized during the treatment process and how they had successfully established baselines and outcomes.

Based on interviews, onsite observation, and review of relevant test data and documentation, a professional high-level map was created, showing each process step, the owner of each process step, and the expected results (Figure 1). Each process step is defined by completion of a specific task or milestone. In addition, a general overall measurement model was created, showing the integration of the measurement tools into the treatment process as they occur (Figure 2). Guenther then cites two landmark HBOT studies, one guided by Dr. Paul G. Harch, M.D., and the second study conducted by a team of distinguished researchers at the Tel-Aviv University for their pioneering work and protocols that the Rocky Mountain Hyperbaric Institute now incorporates. [1,2]

Ownership

Guenther observed that staff interaction was characterized by collaborative leadership and complimentary practices. Staff development aligns around the vision and core values of safety first, availability, and affordability, with Ryan and Eddie leading through their positive examples. Education and coaching occur daily on the job in real-time situations, in building teamwork, and in integration of both TBI and PTSD treatment. A key person in that integration process is Pepe Ramirez, whose role as the chief PTSD counselor adds value by way of bringing his collective experience and training acumen to the clinic on a daily basis.

Dr. Julie Stapleton, the medical director, anchors the overall process, providing medical research, diagnostic skills, and an oversight of the entire process of patient healing. In summary, Guenther validates that the Rocky Mountain Hyperbaric Institute provides a workable business model that offers best practices and one that utilizes the proven treatment protocols and measurement methodologies initially pioneered by Dr. Paul Harch and Tel Aviv University.

PROCESS MAP: Rocky Mountain Hyperbaric Institute

Process Steps	Purpose and Process Step Owner	Results
Veteran Discovery	Veteran learns about Rocky Mountain Hyperbaric Institute via web, Facebook, word-of-mouth, referral, or other means.	Veteran submits application
Exam Screen	Owner: Dr. Julie Stapleton, MD • Screen for chamber fitness. • Identify criteria for success. • Prescribe treatment.	Prescribed Treatment
Orientation and Pre-Assessment	Owner: Eddie Gomez, CHS • General orientation to testing and measurement. • Test administration and interpretation. • Supporting documentation for tests. • The rights and responsibilities of test takers.	Completed Testing and Measurement
Treatment	Owner: Ryan Fullmer, CHS • General orientation to treatment protocol. • Conduct the HBOT protocol to include 40 treatment sessions (5 days/week), 60 minutes each, with 100% oxygen at 1.5 ATA. • Safety.	Completed Treatment
Integrating the PTSD Program	Owner: Pepe Ramirez, Counselor • Schedule counseling in conjunction with clinic's initial interview to occur parallel to HBOT treatment. • Integrate counseling with physical exercise regimen.	Completed Counseling
Post-Assessment	Owner: Eddie Gomez, CHS • Observe behavioral responses to treatment. • Retest and track results.	Completed Post-Assessment
Monitoring	All principals: Continue to assess extent to which ongoing innovations and improvements help make services more safe, available, and affordable.	Periodic Follow-up

**Figure 1. Process Map created by Dan Guenther,
Veteran Advocate © 2015**

Process Map of HBOT Measurement Tools

Measurement Tools	Referral/Diagnosis	Pre-Assessment	Treatment	Post-Assessment
Qualitative Data-Gathering Methods	Counseling services, Medical diagnosis and prescription for treatment, Document Review, Observation, Interviews	Document Review, Observation, Interviews, Administration of the Rivermead Post-Concussion Symptoms Questionnaire, PCL-M, PHQ-9, PqoL, MAST, and DUST	Interviews, Observation	Document Review, Observation, Interviews, Administration of the Rivermead Post-Concussion Symptoms Questionnaire, PCL-M, PHQ-9, PqoL, MAST, and DUST
Quantitative Data-Gathering Methods	Medical diagnosis and prescription for treatment, Commercially available and validated tests	Administration of commercially available and validated tests: CNS Vital Sign Report with Domain Dashboard for 14 areas including validity indicators	Medically-related monitoring and measurement per application of the HBOT protocol during treatment sessions	Administration of commercially available and validated tests: CNS Vital Sign Report with Domain Dashboard for 14 areas including validity indicators
Multi-Method Assessment/Research Applications	Counseling services, Medical diagnosis and prescription for treatment, Process/Relationship Mapping	Process/Relationship Mapping, individual testing and setting a baseline for the tracking average percentile increase of improvement indicators	Process/Relationship Mapping,	Process/Relationship Mapping, tracking average percentile increase of improvement indicators

Figure 2. Measurement Process Map created by Dan Guenther,
Veteran Advocate © 2015

Assessing Brain Function

The clinic utilizes commercially available and validated tests throughout the treatment process (see Figure 2). The tests used are administered before and after the patient's treatment and upon completing therapy. These paper and pencil and computer tests identify cognitive deficits resulting from concussions and/or similar trauma-related injuries. By assessing various aspects of brain functioning, such as short-term memory, attention, concentration, visual spatial capacity, information processing speed, and reaction time, the tests are useful in describing the degree of severity of a particular injury or trauma. This data is crucial to understanding the overall effectiveness of the "Healing Our Heroes" program. Validity is the "truth" of a particular measure, but "reliability" is a test of that measure over time. Collectively, the overall data from the various tests, listed below, indicate dramatic gain on the part of those participating.

Automated Neurol Assessment Metrics (ANAM)

Rivermead Post-Concussion Symptoms Questionnaire

Problems & Complaints List (PCLM)

Patient Health Questionnaire (PHQ9)

Perceived Quality of Life (PQoL)

Drug Abuse Screening Test (DAST)

Michigan Alcohol Screening Test (MAST)

Computerized Neurocognitive StatusVital Signs (CNSVS)

Computerized Neurocognitive Status – Vital Signs (CNSVS)

The above first seven *paper and pencil* tests are administered and recorded on each patient who receives treatment at Rocky Mountain Hyperbaric Institute. The last one, CNSVS, a clinical testing procedure (software) utilized by Rocky Mountain, evaluates, summarizes, and scores in both pre- and post-treatment the neurocognitive state (thinking, reasoning, and memory) of each patient undergoing TBI-PTSD treatment therapy. It is a more sophisticated evaluator.

This software is a cost-effective solution for tracking all veterans and their progress. It is made up of seven basic tests that include: verbal and visual memory, finger tapping, symbol digit coding, the Stroop Test, a test of shifting attention, and the continuous performance test. This all-important computerized tool is valuable in measuring the progress each individual veteran makes as a result of the hyperbaric treatment.

The tool uses well-known, evidence-based medical and health rating scales to assess, track, and score each patient in the various patient profiles shown in Figure 3. For example, the Visual Memory test indicates how well a person recognizes, remembers, and recalls geometric figures. The relevance of this test is related to remembering graphic instructions, navigating, operating machines, recalling images, and remembering an abundance of events. In Figure 3 below, the score of 110 in *Visual Memory* indicates above-average scoring. Similarly, each domain is measured and scored for relevancy for each of the Vital Signs.

Patient Profile:	Percentile Range				>74	25-74	9-24	2-8	<2
	Standard Score Range				>109	90-109	80-89	70-79	<70
Domain Scores	Subject Score	Standard Score	Percentile	VI***	Above	Average	Low Average	Low	Very Low
Neurocognition Index (NCI)	NA	85	16	Yes			X		
Composite Memory	102	103	58	Yes		X			
Verbal Memory	51	93	32	Yes		X			
Visual Memory	51	110	75	Yes	X				
Psychomotor Speed	174	93	32	Yes		X			
Reaction Time*	555	107	68	Yes		X			
Complex Attention*	21	56	1	Yes					X
Cognitive Flexibility	26	63	1	Yes					X
Processing Speed	48	79	8	Yes				X	
Executive Function	34	75	5	Yes				X	
Simple Visual Attention	40	108	70	Yes		X			
Motor Speed	124	105	63	Yes		X			

Figure 3. CNS Vital Signs report example
~ courtesy of CNS Vital Signs www.CNSVS.com

Summary of CNS-VS Assessment Tool

In summary, the data captured before and after treatment is a strong indicator of overall program effectiveness, and, over time, provides a *reliable metric* for benchmarking best practices. Collectively, the overall data from the various tests administered to individual participants indicate a pattern of dramatic gain. Accordingly, the Rocky Mountain Hyperbaric Institute employs the popular CNS-VS report, which offers best practices and a proven treatment protocol from its measurement methodologies. (See Exhibit C for more on CNS-VS reporting software.)

Summary of CereScan Brain 3-D SPECT Scan & Assessment Report

Other options for further quantification and in-depth evaluation of brain injury can be provided by the proven medical team at CereScan. CereScan, with its corporate office in Denver, Colorado, is comprised of experienced professionals who devote their energies to improving quality of life in a wide range of brain-based disorders. This company and its procedures are at the forefront of brain imaging technology. Using captured data from their SPECT imaging procedures, the underlying causes of brain-based disorders can be readily measured and displayed. This technology discloses the presence of a TBI even when other scanning methods do not.

One of the veterans, Colonel George Corbin, USA (Ret) visited Rocky Mountain Hyperbaric Institute twice (two forty-hour sessions) and was successfully treated. He shared his CereScan SPECT & Assessment Report with the authors in order to demonstrate and prove the healing capability of HBOT. The authors are grateful. (See Exhibit D for more on CereScan capabilities.)

In summary, the data captured before and after treatment by SPECT imaging is a powerful option available to Rocky Mountain Hyperbaric Institute's procedures and program effectiveness, which also helps benchmark best practices. Collectively, the overall data from the tests administered to individual participants indicate patterns of meaningful gain. Accordingly, the Rocky Mountain Hyperbaric Institute provides a workable business model, one that utilizes the proven treatment protocols and measurement methodologies.

★ ★ ★ ★ ★

HEALING OUR HEROES
SIGNIFICANT SUPPORTERS

Caleb Gates – Benefactor, Advisor and HBOT Supporter

Caleb Gates, Benefactor and Advisor
~ courtesy of Caleb Gates

At a time when the Rocky Mountain Hyperbaric clinic needed real help, Caleb Gates provided the initial monies to fund the treatment for traumatic brain injured veterans and then move the fledgling clinic to its current professional location. A veteran himself, Caleb joined ROTC at Princeton University and graduated as an Army 2nd Lieutenant in 1953. He was ordered to Fort Sill for artillery training, but he and three pals decided to "have fun" and go to airborne training at Fort Benning, Georgia. Caleb passed the physical exam. His two buddies did not pass. He was then ordered to the 82nd Airborne Division at Fort Bragg, North Carolina, where he served as a Artillery Battery Officer until his discharge in 1955.

Today Caleb is a retired investment banker, having spent forty-eight years in analytical and portfolio management. His career included positions in the Trust Department at First National Bank of Denver, where he managed the monies of high net-worth individuals. In 1983, he was a founding partner and head of Wealth Management of Denver Investment Advisors, now known as Denver Investments. Today it is one of the nation's preeminent advisory firms to institutional and private investors.

From 1969 to 1974, Caleb served on the Board of Director for Children's Hospital Colorado and was one of the founding trustees of its foundation in 1978 that managed to raise $273,000,000. He also served as Chairman of the Board and Chairman of the Nominating Committee. Other involvements of his included the Community Homemakers, the United Way, the Four Mile House Investment Committee, the 10th Mountain, the Kent School Investment Committee, the 10th Mountain Hut Association Endowment Fund, and the Alliance for Choice in Education (ACE).

In 2009, close friend Martin Hoffman, former Secretary of the Army, 1975-77, introduced him to HBOT and the program at Rocky Mountain Hyperbaric Institute, where he began his active support of their Healing Our Heroes 501c3 nonprofit fund, which depended on donor funds to treat and heal Iraq and Afghanistan combat veterans who suffered from TBIs and PTSD. In addition to his own personal financial support, Caleb and his good friend, Jim Wallace were instrumental in establishing major funding support from some of Denver's most prominent citizens. Today, thanks to Caleb, all veterans who enter the clinic receive free forty-hour HBOT treatment and out-of-town veterans are boarded free at the suburban home near the clinic. The RMHBOT Association owes a great debt of gratitude to Caleb!

★ ★ ★ ★ ★

Jim Wallace and General John R. Allen, USMC
~ courtesy of Rocky Mountain Hyperbaric Association for Brain Injuries

Jim Wallace, a Marine veteran and true believer in HBOT, joined Caleb Gates to organize galas and Denver fundraising for the clinic's nonprofit fund. His support has been so financially beneficial that veterans suffering from TBIs and post-traumatic stress will be treated and healed free of charge through the year 2017.

A Texan by birth, Jim was a natural for the oil industry when he and Ray Brownlie became partners in Abilene, Texas, in the 1950s, specializing in lease management and contract negotiations. Then a third partner, Joe Bander, joined them in 1973 and BWAB was formed. They concentrated on the Rocky Mountain and Denver Basin, the Powder River Basin, and a very productive Willison Basin, where ninety producing wells were discovered.

By 1980, the team had explored and discovered oil in the Anschutz Ranch East Unit, Utah, which contains 400 million barrels. They drilled 350 wells and 200 produced results with over 13 million barrels of oil and 44 billion cubic feet of gas. BWAB continues to focus on the Rocky Mountain region and supports ongoing exploratory opportunities throughout the region.

Jim's former association with the United States Marine Corps General John R. Allen attracted the former U. S. and NATO Force Commander and special envoy to the Middle East to speak at an April 1, 2015 fundraising event on behalf of the Rocky Mountain Hyperbaric Association for Brain Injuries. We are extremely grateful to have such distinguished benefactors as Jim Wallace and Caleb Gates.

★ ★ ★ ★ ★

Martin Richard Hoffmann – Advisor and HBOT Supporter

"Marty" Hoffmann, former Secretary of the Army, 1975-1977, tireless Veteran's Advocate
~ courtesy of Martin R. Hoffmann

In Memoriam – The Honorable
Martin Richard Hoffmann: 1932-2014

To know and work with Marty was a unique personal experience. This intelligent, dynamic man graduated from Princeton in 1954, enlisted in the Army, and became a Field Artillery Officer in 1955. He served in RCT 187 and the 101st Airborne Division. After graduating from the University of Virginia Law School, he initially served as a law clerk in the U.S. Court of Appeals. He went on to fill other positions as the Assistant U.S. Attorney, Minority Counsel for the House Judiciary Committee, Legal Assistant to Senator Charles Percy, and General Counsel for the Atomic Energy Commission. In 1973, he became Special Assistant to the Secretary of Defense and the General Counsel for the Department of Defense prior to his appointment as the Secretary of the Army, 1975-1977 under President Gerald Ford.

Marty entered private practice in Washington, D.C., and eventually became General Counsel for Digital Equipment Corporation. He served on the Board of E Systems and as a trustee for the Association of the United States Army. His work on the Secretary of Defense Commission on Base Realignment and Closure was commended for its clarity and timely execution.

However, it was Marty's passion to find an effective treatment for the thousands of brain-injured veterans returning from Operation Desert Storm and the Iraq and Afghanistan wars. As an advisor to Secretary of Defense Donald Rumsfeld, he lobbied for the recognition and treatment of TBIs and PTSD.

His association with hyperbaric oxygen treatment pioneer Dr. Paul Harch led to the first successful treatment therapy of veterans suffering TBIs. Marty's pro-bono efforts found him in congressional offices and Department of Defense and VA headquarters, where he sought funding and support for FDA-approved TBI test trials that would improve the effectiveness of hyperbaric oxygen treatment therapy.

In conjunction with close friend and HBOT advocate Rob Beckman, he witnessed the rapid increase of military suicides, especially by those combat veterans who suffered the signature wounds of the Iraq and Afghanistan wars. "Treat now!" became their motto and mission in life.

To the end of his life, Marty championed HBOT and TreatNOW (www.treatnow.org). He sought out a grassroots start. Then he envisioned that we would create great momentum to treat the veterans with this unique therapy since we now had the means to do so. His wife "Muggy" and his children know how much we honor and miss our Marty.

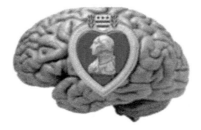

TREAT NOW

The following description of this critical program is testimonial to Marty's leadership and his exceptional work:

TreatNOW: A Citizen Response to the Suicide and Concussion Crises
TreatNOW Mission Statement
Identify and treat veterans and others suffering from concussion/TBI/PTSD.

TreatNOW Message

The TreatNOW Coalition is a group of citizens, veterans, institutions, and service organizations working pro bono to get help to our brain injured troops and others suffering from concussions, traumatic brain injury (TBI), and post-traumatic stress disorder (PTSD). We have already helped treat and heal over 1,400 brain injured individuals.

We are in the midst of suicide and concussion-epidemics: 22-plus service members a day commit suicide, 8,000 a year. Another 45 a day try and fail, another 16,000 plus a year. The CDC estimates between 1.6 million to 3.8 million concussions occur each year. What TreatNOW is proposing will make a significant contribution to ending those suicides and treating and helping to heal brain injuries. Our efforts are reducing risk, health care costs for the country, our wounded warriors and others with brain injuries. And what we are doing is already restoring lives, hope and productivity to a small slice of deserving veterans. This site will brief you on our work. We hope to get you support for the hundreds of thousands who haven't even been told there is treatment and hope.

There is an effective, ethical, medically safe treatment available for TBI/PTSD-wounded personnel and Veterans, but it is not being researched or developed aggressively nor is it being made available to our injured troops or civilians. We are accumulating scientific and clinical evidence of the efficacy of Hyperbaric Oxygen Therapy (See www.hbot.com). It is the only currently available, demonstrably promising treatment for TBI/PTSD and concussions. The Treatment is showing promise for anyone with brain injury, no matter how the injury is acquired.

Our U.S. Military Volunteers who are injured on the battlefield in the line of duty deserve the best treatment our nation can offer. Right now our soldiers suffering TBI and PTSD are prescribed symptom-reducing drugs; in essence, they are warehoused and then discharged, dependent on costly anti-depressants and other anti-psychotic medicines that promote dangerous dependencies and may even result in lethal interactions. Certainly the rising incident of suicides among our veterans is a disturbing indicator that such treatments are not sufficient.

At the same time, TBI is one of the leading causes of morbidity and mortality in the U.S., accounting for approximately 2 million emergency room visits, 230,000 to 500,000 hospital admissions, and 52,000 deaths annually in the United States. Every year, there are approximately 80,000 additional casualties who will be forced to live with significant, and usually permanent disabilities as a result of their traumatic brain injuries, yielding a total estimation of 5.3 million survivors-a number that continues to grow. While such disabilities can be physical, they are often psychological as well. Evidence consistently indicates that survivors of TBI are at increased risk for the development of severe, long-term psychiatric disorders, particularly depression, generalized anxiety disorder, and post-traumatic stress.

Robert B. Alvarez – Veteran Advocate/Therapist/ Counselor and HBOT Supporter

Robert B. Alvarez
~ courtesy of Grady Birdsong, Veteran Advocate

Robert Brian Alvarez, born of Cuban-American parents in New York City in 1959, moved with his family to Miami, Florida, in 1964. Growing up in the inner-city of Miami in the late sixties proved to be a challenging period of depressed economic times with great racial strife for Robert. Amidst these trying times, Robert found the wherewithal and time to become an All City, All County football player at Miami Carol City Senior High School.

Robert found himself in the Marines after high school as an avionics technician, spending most of his enlistment in Jacksonville, Florida. Attending night college classes, Robert completed his military service with a degree in industrial education and returned to Miami, where he began teaching in the same high school he had attended as a teen. Taking on additional responsibilities, he began coaching football, baseball, wrestling, and soccer.

While there, Robert returned to graduate school, earning additional certifications, and found himself in the Special Programs Division of his Tampa, Florida school system. He later was selected to manage the county's new federally-funded program assessing county jail inmates and working with the Sheriff's office in rehabilitation and probation. This program won state recognition and became Florida's new drug courts model. After completing his master's degree in rehabilitation counseling at University of South Florida, Robert began a new career with the Florida Department of Corrections, evaluating and counseling prison inmates.

After moving to Colorado in 1997, Robert ended up working for El Paso County District Attorney, Juvenile Division in Colorado Springs, Colorado. During the "surge" and an escalating war in Iraq, in January 2007, Robert began counseling Fort Carson transitioning and wounded soldiers. He also began a part-time private practice treating convicted sex offenders. Showing initiative and spirit at Fort Carson, he was quickly selected as the first counselor to a newly-designed pilot program, the Warrior Transition Unit (WTU). This program was specifically designed to assist moderate and severely wounded soldiers. This experience led to his assignment as the National Organization's Counselor to the U.S. Army Wounded Warriors pilot program. In 2009, Robert chaired a successful effort to implement the 6th National Veterans Court in Colorado Springs near five military bases.

Robert's affiliation with HBOT began after finding out about Sergeant Marguax Mange's treatment at the Harch clinic in New Orleans. She came away from there almost healed and with new hope. Once back home in Denver,

Robert found another clinic in the Boulder area and got her a second session of "dives" scheduled. It was all uphill after that for young Margaux.

Robert has nothing but praise for the Rocky Mountain Hyperbaric Institute and their treatment of Margaux. He is quick to tell all, "I now think HBOT is one of the most amazing medical treatments that I have ever seen in the treatment of combat veterans with concussive injuries. I have seen nothing like it to date ... and I have been working with wounded veterans for about eight or nine years now. It is a phenomenal treatment. The side effects are minimal. I have sent over fifteen veterans for HBOT treatment. I have witnessed nothing short of miraculous recovery with all of those people."

He goes on to tell of the current obstacles, "Until insurance or the military embraces HBOT for these kids, we will see no real progress in concussive treatment. Veterans are living on fixed incomes and have very limited resources ... they themselves cannot afford it. I cannot say enough about it ... it is not the only treatment out there, but in my opinion it is absolutely the most important and the most successful of any of the others."

In 2013, he and Andrew Pogany created the first nonprofit veteran service organization aimed at preventing the wrongful discharging of the nation's wounded soldiers, entitled the Uniformed Services Justice and Advocacy Group (USJAG). Working with active duty service members, Robert investigates discharges and prevents wrongful discharges of hundreds of wounded combat veterans while maintaining his life as a single father of two teenage boys and serving as a community volunteer. He is also a life member of the Marine Corps league, a martial artist, an avid motorcyclist, and a therapist in private practice of Colorado's state-approved sex offender program.

Pat Smith – American Legion Veteran Advocate and HBOT Supporter

Pat Smith, American Legion Colorado Adjutant, The Department of Colorado
~ courtesy of Darrell Meyers, American Legion *Observer* Editor

Charles Pat Smith began his affiliation with the military in 1965, when he joined the United States Army. He completed basic training and emerged into Advanced Infantry Training and on to a Military Occupational Status (MOS) of financial specialist for the Army. He readily adapted to an artillery unit at Fort Sill, Oklahoma. After advanced training, Pat spent a short time there and finally received his orders for the Republic of Vietnam in the spring of 1965. As a fresh new soldier arriving in-country, it was determined that he would be assigned to the Headquarters United States Army strength accounting section at Tan Son Nhut Air Base. Needs at the time were for accounting rather than finance people. The mission of his new unit and his new job was to accurately affirm all personnel in-country, including hospitalized, killed-in-action, wounded-in-action, prisoner-of-war, and missing-in-action persons, while keeping a running count of all people coming and going.

During his time at Tan Son Nhut and working at this special accounting unit, Specialist Smith was selected by his commanding officer to assist in testifying to the accuracy of the military personnel numbers being kept by their unit section. At that same time, President Lyndon B. Johnson had hastily dispatched Secretary of State Robert McNamara to clear up the troop strength number discrepancy being reported in the news media to the American people. In attendance at the conference were General William Westmoreland, several of his generals and admirals, assorted United States senators, congressmen, and Specialist E5 Charles Pat Smith, along with his commanding officer and other supporting specialists. At the end of the meeting Pat recalls Secretary of State McNamara bluntly stating that, in effect, "Gentlemen, we need to keep this war going because it is good for the economy back home." At the time this comment did not have an impact but would later.

April of the same year, Captain Robert Weskamp, shot down and captured over North Vietnam, was entered into prisoner-of-war status, which immediately caught Specialist E5 Smith's attention. Weskamp happened to be Pat's best friend from early grade school through high school. This was an attention-getter and a real dilemna for a young soldier.

Fast forward to Pat's next assignment at HQUSAJ, Pacific Theatre Officers records unit, where he dutifully reconciled all flag officers' "201 Personnel Files" (which contains all of their personal military records). Pat was nominated Soldier of the Quarter in that command, which was a special honor. Additionally, he was selected for the Soldier of the Year competition and was summarily eliminated by the simple question of who had won the first Super Bowl in 1968. Smith, approaching his expiration of term of service (ETS) was offered a promotion to E-6 if he would re-enlist. With a clear vision of where the war was going, more maturity, and the memory of Secretary McNamara's parting words in 1966, he refused the offer.

Pat returned to civilian life and found the "voice" he needed to talk about his misgivings with the way the Vietnam War had been waged utilizing young men like himself. In 1973, Pat became the first Vietnam Veteran Post Commander in the State of Colorado. All of them in the past had been WWII or Korean veterans. In 1980, he became the first Vietnam Veteran State Commander for Colorado and tried for the candidacy of National Commander of the American Legion. His father's untimely death put the National Commander run on hold.

The year 1979 was a banner year for Pat. The American Legion of Colorado rendered him as Colorado's Outstanding Vietnam Veteran and sent him to visit President Jimmy Carter at the White House. Later in 1981, while serving as State Commander, he was selected along with seven other American Legion Vietnam Veterans to meet President Ronald Reagan in order to brief him on the POW/MIA issues and a proposal that was being debated in Congress to relocate the vet centers back into the hospitals. As a result of this briefing, President Reagan made the POW/MIA situation a top priority in his administration. Additionally, the president went to Congress and asked that the number of standalone vet centers be increased, which happened afterwards.

Charles Pat Smith's dedication and American Legion service to the American servicemen of all branches rendered him a very distinct personal honor. He was chosen by Governor Richard Lamm as one of two Colorado Vietnam veterans to travel to Washington D. C. for the breaking of ground at the Vietnam Veterans Memorial on the Mall in March 1982. Later, he led the Colorado delegation for the opening ceremonies and dedication of the memorial.

Pat now serves as Colorado Department Adjutant for the Legion and has done so for the past twenty-five years. Pat initiates assistance to needy veterans throughout the State of Colorado and is quick to help when the need is there for our military veterans and their families.

In 2009, Colonel Robert L. Fischer approached Pat and Darrel Meyers, the Legion Public Relations Chair, about supporting "a new treatment for traumatic brain injuries, which is called hyperbaric oxygen treatment." Their first impression of this treatment was that it was *voodoo* medicine. After receiving various reports that provided more detailed information, Pat and Darrell engaged in more discussion and began a more inclusive investigation into this treatment. They then followed several veterans' progress in the program, with interviews at the beginning, during, and after the treatment sessions and gradually came to see that HBOT is a successful treatment therapy. Ultimately, Pat expressed, "Hey, this treatment works for our kids, let's support it."

Immediately, Pat, Darrell, and Tom Bock, past National Commander, swung into action and were successful in commandeering the National American Legion ruling body to sign on as a supporter of HBOT. Since that time, the American Legion has been at the forefront in fundraising. Pat and Darrell, having written many articles to explain the program to the legion at all levels and to the general public, have continued the vast support network of the American Legion and focused it onto a worthy cause—the "healing of our heroes."

In his own words, Charles Pat Smith presents the following thoughts, "I believe it is the most significant treatment today for traumatic brain injuries and related post-traumatic stress disorder. This treatment gives hope to thousands of veterans that they can return to some state of normalcy and eliminate the mountain of pills prescribed by the VA for treatment. For our government leaders not to recognize the value of HBOT is not only inexcusable but perhaps immoral. The American Legion will continue to fight for medical recognition of this important treatment program and I will be right in the middle of the fight."

Tom Bock – American Legion Veteran Advocate and HBOT Supporter

Thomas L. Bock, Past National Commander 2005-2006, American Legion
~ courtesy of Darrell Meyers, American Legion *Observer* Editor

Tom actively participates in the daily activities of the world's largest wartime veteran's organization in the United States of America. The American

Legion, with a membership of almost three million members, chartered and incorporated by Congress in 1919, is one of the world's most patriotic veteran organizations. The legion actively devotes its energies to helping veterans, mentoring youth, and sponsoring wholesome programs within communities, while advocating patriotism, honor, national security, and continued devotion to service members and veterans.

Tom, a Vietnam era veteran of the United States Air Force, has been a member of the Legion, serving in numerous capacities of leadership both locally and nationally throughout his long career. Traveling extensively to all fifty states and thirteen foreign countries, including Iraq and Guantanamo Bay Cuba while National Commander 2005-2006, Tom conferred with President George W. Bush, members of his cabinet, Secretary of State Condoleezza Rice, and Secretary of Defense Donald Rumsfeld concerning Legion matters. Addionally, Tom sought out and engaged members of Congress in various American Legion-related matters.

He presently supports the State Veterans Nursing Home at the past Fitzsimons Army Hospital location in Aurora, Colorado. Ever active in the United Veterans Committee of Colorado, he served as its President, 1st Vice President, and 2nd Vice President through the years of 1999 to 2003. Presently, Tom serves as Colorado Department Deputy Adjutant and additionally officates as State Director, American Legion Colorado Boys State. He commandeers the yearly High School Oratorical Contest for Colorado Boys State.

Prior to serving as the American Legion National Commander in 2005-2006, Tom became Colorado State Commander in 1992, then went to work on the National Legislative, Veterans Affairs & Rehabilitation, Internal Affairs Commission. Afterwards, he officiated as National Executive Committeeman from 1995-2001and moved to the Foreign Relations Commission, 2003-2005, as their chairman. Retiring from AT&T in Denver, Colorado, after an almost thirty-year management career, Tom remains active with his beloved American Legion, married to his wife Elaine of forty-nine years, and both are the proud parents of four children and six grandchildren.

Tom is committed to the Healing Our Heroes program and readily tells anyone who is interested in this endeavor, "When we first learned about the hyperbaric oxygen treatment therapy being tested in Boulder, we were extremely interested. Helping raise funds and supporting HBOT quickly became our primary mission. It is very rewarding to see the tremendous progress being made in restoring complete lives for these true American heroes."

Darrell Myers – American Legion Veteran Advocate and HBOT Supporter

Darrell Myers, Editor, American Legion Observer
~ courtesy of Darrell Myers

Darrell comes to Colorado from Cedar Rapids, Iowa, graduating from Williamsburg High School in 1963. Darrell enlisted in the United States Army after high school and became a tracked vehicle and wheel specialist, working his way to Sergeant (E-5). Darrell went into the active Army Reserve in 1969 and eventually served as a maintenance supervisor for the 244th Engineers and attained the rank of Sergeant First Class (E-7).

A graduate of Red Rocks Community College in Denver, Colorado, Darrell took a degree in Industrial Engineering in 1976, prior to an active duty assignment in the regular Army beginning in 1977. While on active duty, Darrell became a Warrant Officer and stayed in the Army until 1982. He left as a Chief Warrant Officer (CWO2) of Motor Maintenance and became a college instructor and program manager at Central Texas College, European Campus, in Germany from 1982 through 1990. In those years, he taught in the automotive and heavy diesel mechanic programs and eventually transitioned into the Military Occupational Status Performance Assistance Program.

Darrell is an ASE Certified Master Automotive Technician and Master Heavy Diesel Technician. Now editor and chief photographer for the American Legion, Department of Colorado, Darrell also functions as their Chairman of Public Relations and Employment. He has won numerous national American Legion awards for the newspaper and himself.

Darrell began supporting hyperbaric oxygen treatment therapy in 2009 by introducing his nationwide American Legion readers to the healing benefits of HBOT through a series of newspaper articles. When asked about HBOT treatment and its success he readily replies, "I am passionate about hyperbaric oxygen treatment because I have seen it work; actually it seems more like a miracle that takes place after only a few dives in the chamber. I have seen these kids unable to speak, think clearly, or display confidence, and finally come out of the treatment smiling, relaxed, looking you straight in the eye, and able to express themselves after all of those prescribed dives. Yes, it is a miracle!"

Sandy and Stan Flint – Volunteer & Advisor and HBOT Supporters

Sandy and Stan Flint
~ courtesy of Stan and Sandy Flint

Sandy Flint, born in 1941, was raised in Seattle, Washington, and moved to California when she was twenty-one years old. Sandy migrated to Colorado and arrived in Denver with just eleven dollars in her pocket, no job, and a room reservation at the YWCA in the 1960s.

True to her self-starting character, she landed a job immediately and never looked back. Sandy succeeded in every venture or endeavor she chose to undertake. Her early career included becoming a dental assistant, a ballerina, a bartender, the vice president of an advertising agency, a radio show hostess, and finally, a realtor. She maintained a successful real estate career that lasted the span of forty-five years.

In 1968, Sandy met Stan Flint. They married in 1970 and have five children and eleven grandchildren. Sandy and Stan donate numerous hours helping many people in need within the military family.

Sandy has an extensive military family of her own. Her father, a decorated Commander, served in the United States Navy the entire duration of WWII. Sandy's uncle was also a U.S. Navy officer during World War II, and her husband, Stan, is a Marine veteran and a career police officer. Her brother is a Marine veteran of the Vietnam War. Sandy's dedication to veterans comes from this family connection to the military and her desire to support veterans who need her help.

Over the years, she began volunteering to help veterans in a variety of programs and nonprofit organizations. She was the treasurer of American Legion Post 11-11 and also was the past-president of the District 5 American Legion Auxiliary for four years. Sandy became a member of the Broomfield Fraternal Order of Eagles.

As Chairman for the American Legion's Colorado Wounded Warriors, Sandy raised thousands of dollars for these warriors and has been able to help many needy veterans. In one case, she was the contact that the fledgling Rocky Mountain Hyperbaric Institute made when Staff Sergeant Dean Sanchez, the USMC Wounded Warrior Representative for five western states including Colorado, made a very special request of Sandy.

It seemed that Sanchez had located a badly wounded Marine veteran living near or under a bridge in Colorado Springs. The Marine had suffered brain concussions in two Marine truck accidents, was partially blind, and had no disability or other support from The Denver VA Hospital. He suffered from obvious TBI as a result of these accidents.

Sandy played a major role in rescuing and supporting this Marine in need. First she "found" funds to pay for his rent in a Colorado Springs apartment, thanks to her American Legion Auxiliary. When he was finally off the streets, Dean Sanchez was able to find a veteran's advocate who supported the new HBOT program at the Rocky Mountain clinic, and Eddie and Ryan agreed to provide a forty-hour treatment program to heal him. Not only was this Marine almost totally healed, with his eyesight and speech restored, but when he first arrived at the clinic he'd had an infected ankle with a five-inch gash that was wrapped in a dirty bandage. To everyone's surprise, his ankle was totally healed, with no sign of infection, after just five hours in the oxygen chamber.

Since 2009, Sandy has been a strong supporter of the Rocky Mountain Hyperbaric Association for Brain Injuries and now serves on their Board of Directors. Sandy is passionate about serving on the board and raising funds to help this great cause. "I really do love it," Sandy has been known to tell others about the clinic and the work being done there. When anyone meets Sandy, they can immediately see that she is truly dedicated to helping and healing our veterans.

Rick Baum – Volunteer & Advisor and HBOT Supporter

Earl Riggins, WWII USS Indianapolis survivor with Rick Baum (right)
~ courtesy of Rick Baum

Rick Baum, a veteran of the United States Marine Corps, served as a past minister and also with the Nebraska Country Schools as Field Service Director and lobbyist. Additionally, he established veteran burial programs throughout Colorado, Nebraska, and Indiana. Rick is now a stanch Veteran Advocate for Healing Our Heroes, lobbying state legislation in both Colorado and Indiana.

His two sons are military veterans. Youngest son, Nathaniel, served multiple tours of the Middle East in the Marines and oldest son, Benjamin, deployed to Iraq as a Nebraska Army National Guard soldier. After returning, son Ben became a police officer and was diagnosed with PTSD by the VA. Ben passed away in 2008, leaving behind a widow, a daughter, a brother, and his parents.

Rick continues his service to the community through the Marine Corps League, U. S. Marine Corps Memorial, Golden, Colorado, the American Legion (past Service Officer & Public Relations Chair), and the American Legion

Motor Cycle Club riders. He continues his Veteran Advocate liaison for the Rocky Mountain Hyperbaric Association for Brain Injuries by sending afflicted veterans to the program from his current home in Indiana.

His love of country and serving as an advocate comes from seeing the effects of the invisible wounds of veterans, losing a son, and serving God. Rick continues to provide educational presentation on TBI-PTSD and reaches out to OEF and OIF veterans suffering from TBIs and post-traumatic stress. He coordinates fundraising while encouraging financial donations to the Healing Our Heroes program.

In his relentless pursuit to help those serving our military, Rick explains what he has observed over the years: "These men and women return home adversely changed by their traumatic experiences. They come home with invisible wounds that haunt them 24/7and their unsuspecting families also suffer vicariously. These stories before treatment speak of fear, anxiety, and horror."

Rick goes on to describe the rescuing effects of hyperbaric oxygen treatment: "They have their lives and families back. Debilitating headaches and prescribed pharmaceuticals have decreased. They are taking responsibility for themselves as their minds and memories become revitalized. They demonstrate more emotional control, and function as husbands, caring fathers, and contributing citizens in the home and community. They seem to have improved lives as a result of this treatment."

Rick speaks of one veteran who motivated his daughter's whole elementary school to rally and collect over $950 through a "Pennies for Patriots" fundraising drive in just three short weeks by collecting coins. He also mentions that in working with veterans in Indiana, their personal testimonies have inspired key retired veterans to pursue state supported legislation to treat "signature wounded" veterans. Likewise, these healed veterans inspire American Legion and VFW Rider Chapters to host fundraising events that promote the hyperbaric treatment message.

Robert L. Beckman, Ph.D., Senior Knowledge Management Professional

Robert L. Beckman, Ph.D., Senior Knowledge Management Professional
~ courtesy of R. Beckman

From the earliest days of the Rocky Mountain Hyperbaric Institute, there have been few supporters like Rob Beckman. A good friend and close associate of Marty Hoffman, Rob has assumed the management and coordination of their **TreatNOW** program. Rob is the Chief Knowledge Officer for the Foundation for the Study of Inflammatory Disease and is the key voice of **TreatNow.org**, which is part of a nationwide coalition focused on ending the suicide epidemic so prevalent with today's military veterans. He is currently responsible for validating a national network of hyperbaric clinics that treat TBIs and post-traumatic stress cases through the architecting of pertinent data collection and analysis platforms for those maladies. His dedication to the research and verification of promising therapies, medical procedures, and documented TBI and PTSD results have provided all advocates of HBOT therapy with the very latest studies and data in treating our veterans.

Rob distributed one of the most important studies that proves HBOT's success: the Israeli study that was conducted and also published in 2013 by a distinguished team of researchers drawn from Tel-Aviv University, The Assaf Harofeh Medical Center, and the Institute of Hyperbaric Medicine;

Zerifin, Israel. It tested and documented the brain's capacity to heal itself, especially when it suffers concussions like combat veterans do. Entitled "Hyperbaric Oxygen Therapy Can Improve Post-Concussion Syndrome Years after Mild Traumatic Brain Injury," and presented as a randomized prospective trial, it not only documents and measures treatment and healing of fifty-seven test trial cases but also refutes the naysayers who focus on placebo effects while ignoring valid HBOT success.

Rob Beckman has been building knowledge management systems most of his professional career, primarily in the intelligence community and also with the Department of Defense. He led efforts in planning and organizational strategies and war room development, and also functions as a data integrator within the counter-intelligence and counterterrorism communities, many of which he is a member.

With an extensive work, academic, and publishing career spanning thirty years, he has owned consulting and software companies. Married with three sons, Rob, a Ph.D. in international relations, is an expert in nuclear non-proliferation and knowledge management. He flew KC-135 aircraft in the United States Air Force and is a Vietnam Veteran who joins us as a Veteran's Advocate supporting HBOT.

When asked why he is so passionate about working with veterans now suffering the signature wounds that lead far too many to suicide, Rob readily refers to a quote by his close friend Martin R. Hoffmann, the former Secretary of the Army from 1975-1977. The Honorable Marty Hoffmann recently had exclaimed, "... the effects of the TBI/PTSD epidemic will linger for years, with its epidemics of suicides and broken homes, if not confronted now. The more than 1,000 HBOT-capable clinics in the U.S. represent a pivotal asset in addressing this, the most recent national tragedy affecting military wartime wounded."

Then Rob also exclaims: "I was privileged to work side by side with Former Army Secretary Martin Hoffmann (RIP, 2014) for four years. He was never more articulate than when he demanded that DoD and the VA take a risk for the wounded. He and many of us felt that the United States government had to be held accountable for Title X mandates that commanders bear in their responsibility for the health, safety, and welfare of the troops, especially the wounded and their families. It is humbling to bring the same urgency to this fight that he so ably embodied. Together, we evolved the theme, Treat Now. Delay is deadly."

Sarah Stoltman – Cara Mae Melton – Clinician's/Veterans Advocates, Norfolk, Virginia

Cara Mae and Sarah in front of their chamber at the Hampton Roads HBOT Clinic
~ courtesy of Sarah Stoltman

In 2013, Ryan Fullmer met Sarah Stoltman, who had extensive HBOT experience. She had also worked for Dr. Paul Harch, in her early years becoming a certified hyperbaric technologist. Along with her business manager, Cara Mae Melton, they opened the Hampton Roads HBOT Clinic in conjunction with the Renova Wellness Center. Since then, Sarah and Cara Mae have become certified hyperbaric specialists and now share their unique healing treatment of the more difficult cases, collaborating as well as sharing insights.

Sarah and Cara Mae have become a "sister clinic" to Rocky Mountain Hyperbaric Association for Brain Injuries, as they too have had great success treating and healing signature wounds of war. Cara Mae's husband is a career Navy Seal. It is now known that at least fifty percent of active duty Seals are diagnosed with TBI and PTSD. His treatment with HBOT at their clinic in Norfolk was highly successful, and he has been assigned to his next duty station.

Sarah has spent her last nine years treating over three thousand patients with HBOT. In recollecting those years she states: "It sounds crazy, but this is my purpose in life. When you see these people, especially our wounded veterans, walk into the clinic, often alone, numb, and hopeless that anything will ever help them, since the military has thrown them to the curb with less than honorable and medical discharges, you want to do whatever you can to help. Then you see them start walking and smiling again for the first time in years as their slurred speech improves and they regain their memory and sleep. Most important, they get off their overdosed medicines that make them suicidal and they begin to mend their lives and return to very happy wives, children, and families. That is what motivates me to do this important work. Like Ryan Fullmer's Rocky Mountain Hyperbaric Institute, we too focus on healing our combat veterans."

Sarah started college after high school and did the "due diligence" that all teenagers struggle with while figuring out what they want to do in life. Taking several courses, she could not figure out what she wanted to do and dabbled with becoming an accountant, or maybe a chef, marine biologist, or even a zoologist, changing her mind frequently. Finally, she came upon a commercial for a diving school and decided to see what it entailed.

As she recalls, the recruiter told her that there were not many women in the field of diving. Sarah remembered those first days, "It was hard physical work ... not having a tough bone in my body, I went ahead and decided to commit to this one-year course of commercial diving. As I remember it, we started with forty-five students in the class and when graduation time came there were only seven who graduated. I hung in and graduated."

After graduation, Sarah sold her car and hitched a ride from San Diego to New Orleans, Louisiana. After finding work in the Gulf of Mexico, Sarah found the work grueling and hard. For the first year-and-a-half of constantly being offshore, fixing leaks on pipelines in zero visibility, she found that the only thing she enjoyed about the job was operating the hyperbaric chambers as the dive medic who treated divers with decompression sickness. It was then that she found a local clinic that used hyperbaric chambers and asked if she could volunteer.

Falling in love with being able to help people in this field really motivated young Sarah. As she tells it, "The changes I saw in patients were phenomenal. No one really knew anything about hyperbaric oxygen treatment therapy at the time. After proving myself as a volunteer, I was offered a full-time position as a hyperbaric technician by one of the pioneers of hyperbaric medicine, Dr. Paul Harch. Working alongside Dr. Harch and Dr. Keith Van Meter in the clinic, I developed a pretty good knowledge base, learning more about the research and the healing process."

It was in Dr. Harch's clinic that Sarah had her first opportunity to come in contact with and treat veterans with TBI and PTSD. Respecting veterans and working with them became very rewarding for her. As she learned more about the medicine and process, she decided to make a move into a private hospital environment to round out her knowledge. Moving to the Washington D.C. area and then North Carolina for a while, she added to her budding knowledge of hyperbaric oxygen treatment therapy.

Sarah describes this new experience, "When I found an opportunity to be part of the launch team for a clinic in North Carolina, which would give me the experience on the business side and back to a private clinic environment, I jumped at the chance, learning all the different aspects of start-ups. It was unfortunate that this business did not make it, even though it provided me with valuable lessons learned."

Moving to the Norfolk, Virginia, area to be close to her fiancé's family, Sarah started working for a small clinic and recounts what happened after the move, "Six months after I had been with Hampton Roads Hyperbaric, I received a call from a Cara Mae Melton. She, a Navy veteran, had encountered a mild traumatic brain injury. She told me she was researching options for herself and had found a clinic. We clicked immediately. After I got to know Cara through her visits to the clinic, I found that she too was just as bothered as I about HBOT not being available for our military veterans who were returning from theatre by the hundreds with TBIs and PTSD. She shared with me that her own husband was a Navy Seal who had already suffered combat concussions and was still on active duty. Our meeting was fortunate for both of us. It was then that we began strategizing how we could work together to make a difference."

It was a fortunate meeting of "the sisters," who then established the Hampton Roads Hyperbaric Clinic. In Cara Mae's words, "When I met Sarah, I was Chief Operating Officer at America's Mighty Warriors, and prior to that, I co-founded and ran a company called CareConscious, an organization dedicated to educating and empowering family caregivers. Before Care-Conscious, I was the COO and VP of Business Development of LawInfo.com (now a Thompson-Reuters company). I had joined LawInfo after leaving the Navy. In short, I had absolutely no experience whatsoever with HBOT. However, the more I learned about this unique treatment and the roadblocks that made it difficult for the average person to get access to it, I was hooked."

Cara Mae found HBOT to be such an amazing adjunctive healing therapy that she knew she wanted to be a part of making it available to all of those who could benefit from the treatment. Cara Mae was offered the job of Chief Operating Officer of Hampton Roads Hyperbaric, and shortly thereafter, she and Sarah had the chance to buy the clinic. Cara recalls, "Though it was a huge risk, we took it. There was no way we were going to let this opportunity go away and further reduce the chance for our community and our military to get the treatment they badly needed and so much deserved. Since meeting Sarah and building our HBOT clinic, we both have made a major difference in the lives of many combat veterans, policemen, and others who have suffered from traumatic brain injuries, post-traumatic stress, and other maladies that this treatment heals."

Hampton Roads HBOT Clinic, recently cited on television, radio, and the print media for their special work and a significant number of success stories in hyperbaric treatment therapy in the Norfolk area, can talk about the many lives and families they have helped to reconstruct. Cara Mae tells all who listen, "And today we can joke that we are two gals ... with a chamber ... on a mission ... we heal veterans!"

AFTERWORD:
THE PATH AHEAD

Xavier A. Figueroa, Ph.D.

The success stories in this book are only a small sample of the many hundreds of service members that have undergone hyperbaric oxygen therapy (HBOT). An equal, or greater, number of civilian TBI-PTSD sufferers have achieved excellent results with HBOT, adding to the evidence for efficacy of this therapy. For the majority of veterans, and many active duty members, HBOT has been the only therapy that has allowed them to feel relief from their symptoms and recover from their injuries. The source of their injuries, the duration of their symptoms, and the treatments they have tried are all varied and unique from person to person. All underwent standard and accepted medical and psychological interventions, but failed to recover or find long-term relief from their symptoms and injuries.

The commonalities of each active duty service member or veteran is that each sufferer had to discover HBOT on their own or rely on a loose-knit network of charities and clinics to learn about treatment with hyperbaric oxygen. The healthcare delivery system in place was, and continues to be, unequipped, unable, or unwilling to make HBOT available to treat the TBI, PCS, or PTSD of these individuals. Beg, borrow, or steal has been the operational status in order to receive treatment using HBOT. It is still the case now, thirteen years after the first clarion call to action was made to help these affected service members and veterans.

Slowly (too slowly) the tides are shifting in the medical arena. Since 2002, when a small group of veterans, active duty service members, and civilian healthcare providers saw the need, they began to explore and explain the properties of HBOT on the brain. The stories, the clinical trials, the research, and the advocacy are pushing back against the prevailing wisdom regarding brain injuries and TBIs. Now, the word is out and charities are sponsoring treatment for past and present service members. More doctors are trying HBOT, low-intensity light therapy, EMDR, and electromagnetic therapies for these brain injuries and learning how to apply them. The pace is slow and driven by goodwill and charity.

But that is galling in the extreme: healthcare is not a charity. It is an earned benefit—and a right for veterans. When therapeutic interventions are denied— or not paid, which amounts to the same—because of ignorance or active blocking, we have a major problem on our hands. When we restrict doctors from prescribing treatments that have been shown to work, we usurp an important right and obligation from physicians: first, do no harm; second, work to promote health.

Not a Silver Bullet ... but Damn Close

HBOT is part of a growing group of alternative therapies that help re-activate and regrow the brain, even years after the injury. HBOT and many other therapies are ignored, unknown by doctors and unreimbursed by insurance carriers. Hyperbaric medicine straddles a unique line: it is a proven medical treatment that has been around for fifty years, but it is considered by many physicians as an alternative therapy for brain injuries. Animal models and clinical trials have reported results that clearly demonstrate the effects of HBOT, and even pressurized room air, on brain injuries and stroke. It is an effective therapy that can initiate and support healing, but it is not a magical or complete solution. HBOT has to be part of an integrated treatment approach in order for the therapy to be effective.

In the case of TBIs, you need to have a team approach and a game plan to ensure that you stack the odds of healing the brain. This means that you need to have psychological and reintegration counseling for the patient and the family. This is something that the Rocky Mountain Hyperbaric Institute has been doing very well.

You need to have a nutritional support program that provides the essential building block to reduce brain inflammation and promote brain repair and regrowth. You also need to have objective measures that can tell you if the therapy is having any effect on the patient. All of this is required to optimize successful treatment.

Yet, with less than optimized conditions, HBOT has been yielding impressive results for TBI/PCS victims. What is truly thrilling about this therapy is that we are just scratching the surface of what can be accomplished. Combining different treatment modalities (light therapy, electromagnetic therapy, EMDR, nutraceuticals) will yield gains that we cannot begin to imagine for those unfortunate enough to experience a TBI and live with PCS (or PTSD). Unfortunately, the current health insurance system makes the bringing in of these therapies difficult and expensive.

Education, Ignorance and Economics

In North America, medical education covers at least four years of basic training for most Medical Doctor, Doctor of Osteopathy, and Naturopathic Physician programs. Advance training may require up to eight additional years, or more, depending on the specialty. Most physician programs mention hyperbaric or aerospace medicine, but stop short of further education. Most physician programs rely on the pharmacological approach to treating disease, and medical gas research is not a big component of the education curriculum.

Overall, approximately 2,000 active physicians in the United States have received certification in hyperbaric or aerospace medicine, making it one of the smallest specialties in medicine. That means that about ninety-eight percent of all physicians in the United States know little of the properties of HBOT and are unaware of the reported benefits associated with its use in brain injuries, let alone in the approved conditions.

When drugs or medical devices are certified for use by the FDA in the U.S., there is a substantial monetary investment from industry to educate physicians and clinicians on its use, during and after medical school. There is also a great deal of commercial "education" in the form of advertisements that appear in TV and radio ("Ask your doctor about ..."). Why certain drugs are prescribed and used has a great deal to do with the physician's training and education, but drug representatives are out there making sure that physicians don't forget about their product. It is a constant campaign to ensure certain drug products remain in the formulary list of insurers. You have an industry in place that knows how to get molecules and proteins approved, manufactured, and distributed for use. They make it very easy for physicians to prescribe, and they can charge quite a steep price for patented drugs. Drug and device companies actively support the research that helps their product gain more use. Hyperbaric manufacturers barely register as a blip in their funding capabilities.

With hyperbaric oxygen, you have a gas. Not a pill or injectable. The technology is well over 250 years old, and oxygen cannot be patented.

The cost of mono-place chambers are in the hundreds of thousands, up to two million for multi-place chambers. The capital investment is high, you need trained technicians to run the chambers, and physicians will only get reimbursed for thirteen to fifteen approved indications. If physicians use it to treat a brain injury or a stroke, for example, they will have to fight, cajole, send in tons of paperwork, and have the claim denied. When doctors use drugs or devices in a way that has not been certified by the FDA, they use it in what is called an off-label fashion.

Physicians are not encouraged to prescribe treatments in an off-label manner, but they are ultimately responsible for the treatment they prescribe to their patients, and it is legal for the physician to do so. For drugs that have been around for decades, doctors learn to use them in new ways to help alleviate symptoms and aid in healing from a disease or injury. The biggest problem is getting insurance plans to reimburse for off-label drug use. Many insurance companies will not pay for a drug that's used in a way that's not listed in the approved drug label.

In cancer treatment, for example, these issues have been resolved through a 1993 federal legislation requiring insurance (CMS) to cover medically appropriate cancer therapies. This law includes off-label uses, if the treatment has been tested in careful research studies and written up in well-respected drug reference books or medical journals. In 2008, Medicare rules were changed to cover more off-label uses of cancer treatment drugs. But, here is the catch: What qualifies as "careful research studies"?

When it comes to HBOT, insurers will not pay for off-label treatment on the grounds that it is "investigational" or "experimental." There are plenty of medical groups that are willing to counter the results from case reports and clinical trials with HBOT. They don't believe that HBOT works and work hard to ensure that it won't work. Furthermore, there is no lobby, industry group, or federal research group that can promote the interests, or fund research or education for HBOT. If there are, they are doing a piss-poor job. Hence, a "medical/community" insurgency has arisen to try and get HBOT and alternative therapies approved for use.

With HBOT, the economics do not allow for off-label use, especially when you have to fight an insurer who can bring in an "expert" to counter your claim that research supports the use for a non-covered indication. It's either out-of-pocket, charity, or watching a patient spiral down. As a hyperbaric physician or a clinic owner, you are on your own.

The Emerging Treatments

HBOT has been around for over fifty years. It is a safe therapy with few side effects, when safety protocols are followed. Although the list of reimbursable conditions is short, the list of non-covered conditions that it can treat, and may be able to treat, is very long. Books in the U.S. (*The Textbook of Hyperbaric Medicine*, Dr. K.K. Jain; *Hyperbaric Oxygen Therapy*, Dr. Richard Neubauer & Dr. Norton Walker; *The Oxygen Revolution*, Dr. Paul Harch) and the UK (*Oxygen and the Brain*; Dr. Peter James) have been written that detail how it works and how many other applications in disease and injury HBOT can be applied to heal. Studies sponsored by the DoD/VA/Army (*Ref. 1-3*) and civilian reports (*Ref. 4-6*) demonstrate that HBOT is working, far better than the best treatment afforded our men and women in uniform. But, as detailed above, acceptance and use of newer therapies awaits the same fate as HBOT. Becoming proactive is the only way to accelerate the use of these safe and effective alternatives.

Light Therapy (Laser, LED, IR, NIR and Blue)

Low intensity light therapy or low level light therapy (LILT/LLLT) is not as new as you would believe. Research started in the early twentieth century demonstrated that photons (light particles) could affect the biology of cells that did not have specific receptors to capture light energy (like the rods and cones in the eye). The effects of infrared (IR), near infrared (NIR), and blue light have shown the ability to improve wound healing rates. IR and NIR have shown protective effects in mice against Parkinson's disease and (surprise, surprise) brain injuries. The wavelengths are critically important for IR and NIR to function effectively. Not just any old photon of IR/NIR will work!!

Although the clinical data is scattered, the use of this technology has yielded very good results in early animal models (*Ref. 7-12*) and early pilot work (*Ref. 7, 13*). A study sponsored by Cerehealth Corp. in Colorado ("Traumatic Brain Injury in Veterans and Near-Infrared Phototherapy," NCT02635516) was recently finished that looked at brain blood flow in veterans with PCS. The Massachusetts General Hospital is recruiting for an acute TBI clinical trial for brain-injured individuals using LED NIR light ("Low Level Light Therapy (LLLT) with Near Infrared Light Emitting Diodes in Patients with Moderate Traumatic Brain Injury (TBI)," NCT02233413). Both studies should be reporting soon, and the preliminary reports suggest a considerable effect for NIR on blood flow improvement.

The industry for light-based medical technologies is growing rapidly, but there has been no adoption of this technology by insurers for TBI, and reimbursement is spotty for other conditions. Insurers will need to be convinced of the utility of this therapy, but considerable resistance remains. Although there is less historical and cultural baggage association with LILT/LLLT than with HBOT, the climb will be uphill.

Nutraceuticals

This area of emerging research is focused on the use of nutritional supplements and key dietary compounds to promote nervous system regeneration. A key dietary compound that has shown impressive properties has been the so-called Omega-3 fatty acids. In animal studies of TBI, the fatty acids docosahexaenoic acid (DHA) and eicosapentaenoic acid (EPA) have been reported to have the ability to improve neurological function (*Ref. 14-17*). Clinical evidence demonstrates the properties of the Omega-3 fatty acids in restoring and reversing a number of neurological conditions (*Ref. 18, 19*), including a case report for severe TBI (*Ref. 20*). Omega-3 is not the only supplement (*Ref. 21, 22*) to show preliminary and clinical evidence to aid in TBI recovery. Although this is not considered as a drug, it is an inexpensive way to proactively approach TBI treatment. Other foods and supplements are being tested that may have protective and restorative effects.

EMDR

I won't talk about this therapy at length, but EMDR (Eye Movement, Desensitization, and Reprocessing) is the type of scientific discovery that is made when a well-prepared and open mind follows up on an observation. Like HBOT, EMDR is not a magic bullet, but is a neurological and physiological mechanism that helps to heal and needs to be integrated into a psychological and trauma protocol. It has worked exceedingly well with PTSD. The data in EMDR is growing and accepted as a treatment modality by several insurers (including VA and Tricare), yet debate rages on in the medical community.

Electromagnetic Pulse Therapy

The arena of electromagnetic pulse therapy (non-light based systems) relies on direct electrical or magnetic scalp stimulation. Low Energy Neurofeedback System (LENS) (*Ref. 23, 24*), magnetic resonance therapy (MRT), and transcranial magnetic stimulation (TMS) (*Ref. 25, 26*) are just three of a growing number of transcranial stimulation devices that have good preliminary evidence and safety data. Again, none is covered by insurers.

What is Needed?

The current situation within TBI treatment is not an issue of evidence-based medicine. Plenty of research demonstrates that HBOT, low-intensity light therapy, nutraceuticals, and other "alternative" technologies do far better than the standard approaches used in TBI recovery. This is about money, politics, and who is willing to push for new indications for these technologies (and train the doctors).

Hospitals won't lead the charge or offer HBOT for TBI, because there is no reimbursement channel and they are mired in very complex and time-consuming bureaucracy. Non-hospital based clinics can offer treatment, but they have limited budgets, work on thin margins, and are not major financial players. Clinics like Rocky Mountain Hyperbaric Institute, which offer treatments for free or at a steep discount (they offset the cost with charitable donations), do provide an important service, but they are few and far between.

The reality in this medical arena is for the non-hospital, independent-based clinics to take up the charge. They have the freedom to make decisions about what they can offer to patients, and there is a real economic incentive for them. Unfortunately, they are spread out across fifty states, work independently, and do not have the political or economic power to push for new indications. Because they work alone and are marginalized, there is little each clinic working alone can accomplish. Here is where we need to take a lesson from some *Guerilla Grunts*: apply insurgency tactics to the situation. This worked successfully for the civil rights movement and the American Revolution.

There are approximately 1,300 independent clinics in the United States. Many see each other as competition for a small pool of patients in their community, but they provide a service that other medical specialties cannot. If they could be reimbursed by insurance for TBI (or stroke), they would have more business than they could handle, making collaboration between clinics essential.

This also applies for the newer therapies that are currently trying to gain traction and acceptance. Many, if not all, of the "alternative" technologies can and should be used with HBOT where deemed appropriate. Many of the independent clinics incorporate more than one of these technologies to help accelerate the repair of an injured nervous system and show some amazing outcomes. It should be a "no-brainer" for these new technologies and approaches to be used by a larger set of medical practitioners, especially when the technologies are safe and well tolerated.

Unfortunately, there is initial resistance to new ideas. It doesn't matter which field you are in (ask Col. Robert Fischer, Grady Birdsong, Dr. Paul Harch, Orville and Wilbur Wright, and Dr. Barry Marshall[1]), new ideas are received by a community with all the pleasure and willingness of a dead rat at your doorstep.

Resistances to new ideas are not due to the validity or utility of the idea or concept, but by the experience and comfort that the community feels towards those ideas. Ideas take time to become established and accepted, but they need some form of support and active promotion to make sure they don't die before their time.

Hence, the need to establish an insurgency organization that can help unite all these clinics into an effective political and economic block at every level in the United States. Right now, each independent healthcare clinic that provides access to HBOT and other therapies works alone. Each clinic can speak to insurance boards, state or federal representative offices, and other officials until they are blue in the face. Until a national organization pools and coordinates the activities of these clinics, and those of the equipment manufacturers, there will be very little action.

Each clinic, at the state level, needs to address the legislature on the benefits of using HBOT and the newer technologies. This needs to be coupled with active negotiation at the national and interstate level with insurers. Standards have to be set for policy and protocol. Lastly, there needs to be a national database (a registry) to share the clinical results from the use of HBOT and allied technologies. Although not the "gold standard" accepted by academic medical journals, a registry allows for outcome data to be pooled and the development of best practices to emerge. It also provides a direct link to demonstrate to insurers that therapies do or do not work. How many treated patients in the U.S. have clear results, in the form of pre- and post-HBOT scores, sitting in a clinic office? You can call several dozen reports or testimonials as anecdotal, but when the numbers start reaching into the hundreds or thousands, you run into the realm of the plausible and real. This is where we need to ensure that the data does not go away, because it serves a much greater goal in our lives—ensuring that we are able to take advantage of effective tools to heal.

As stated previously, the problem with the reimbursement of certain medical and alternative therapies is not due to the lack of scientific or clinical evidence. Insurers only cover indications when they are "reminded" of their responsibility to their policyholders. Practitioners play an integral part in making sure that new medical advances are applied and covered by the insurer. This needs to be a coordinated effort, because there is strength in numbers, facts, and the reality of what a small band of dedicated individuals can accomplish.

Alternatives exist for TBI and PTSD, but a concerted effort in educating physicians and clinicians of the reality of these therapies must be made. Insurers need to be made aware of the benefits to their policyholders (and their bottom-line) for including these therapies in their formularies. The time for action is now, otherwise many service members and veterans will continue to face a hell that is fueled by medical ignorance and political apathy. The miracle workers in South Boulder Road are only the tip of the iceberg to treat brain injuries.

Xavier Figueroa serves as a Director and Board Member of the National Hyperbaric Association, a co-founder and president of the Brain Health & Healing Foundation, and former Director of Scientific Research at the Restorix Research Institute. Dr. Figueroa has been performing neurological clinical research since 1995 in the field of Alzheimer's research, as well as basic research in neuron biology, cancer research, bioengineering, and the biophysics of water in cells.

Xavier received his doctoral degree in 2003 from the University of Washington's Program in Neurobiology & Behavior in the Department of Environmental Health and Toxicology. His doctoral training was followed up by two post-doctoral fellowships within the Department of Bioengineering. His experience within academic research exposed him to the undercurrents of research politics, both local and national, that silently shape how science is done and how the results are interpreted. This experience made him aware of the need to maintain an open mind and champion deserving and fruitful ideas, such as the use of HBOT and allied technologies for the treatment of neurological injuries.

Xavier is currently the owner and Director of Clinical and Scientific Research of Sciencia Incognita Consulting LLC, a consulting and management firm. Sciencia Incognita focuses on providing expert scientific advice on specialty areas of biology, biotechnology, and neurobiology, and design of clinical trials and basic research experiments.

Afterword References

1. Miller RS, Weaver LK, Bahraini N, Churchill S, Price RC, Skiba V, Caviness J, Mooney S, Hetzell B, Liu J, Deru K, Ricciardi R, Fracisco S, Close NC, Surrett GW, Bartos C, Ryan M, Brenner LA. Effects of Hyperbaric Oxygen on Symptoms and Quality of Life Among Service Members With Persistent Postconcussion Symptoms: A Randomized Clinical Trial. JAMA Intern Med. 2014. Epub 2014/11/18. doi: 1935931 [pii] 10.1001/jamainternmed.2014.5479. PubMed PMID: 25401463.

2. Wolf G, Cifu D, Baugh L, Carne W, Profenna L. The effect of hyperbaric oxygen on symptoms after mild traumatic brain injury. J Neurotrauma. 2012;29(17):2606-12. Epub 2012/10/04. doi: 10.1089/neu.2012.2549. PubMed PMID: 23031217.

3. Cifu DX, Walker WC, West SL, Hart BB, Franke LM, Sima A, Graham CW, Carne W. Hyperbaric oxygen for blast-related postconcussion syndrome: Three-month outcomes. Ann Neurol. 2014;75(2):277-86. Epub 2013/11/21. doi: 10.1002/ana.24067. PubMed PMID: 24255008.

4. Efrati S, Ben-Jacob E. Reflections on the neurotherapeutic effects of hyperbaric oxygen. Expert Rev Neurother. 2014;14(3):233-6. Epub 2014/01/30. doi: 10.1586/14737175.2014.884928. PubMed PMID: 24471697.

5. Boussi-Gross R, Golan H, Fishlev G, Bechor Y, Volkov O, Bergan J, Friedman M, Hoofien D, Shlamkovitch N, Ben-Jacob E, Efrati S. Hyperbaric Oxygen Therapy Can Improve Post Concussion Syndrome Years after Mild Traumatic Brain Injury - Randomized Prospective Trial. PLoS One. 2013;8(11):e79995. doi: 10.1371/journal.pone.0079995.

6. Harch PG, Andrews SR, Fogarty EF, Amen D, Pezzullo JC, Lucarini J, Aubrey C, Taylor DV, Staab PK, Van Meter KW. A phase I study of low-pressure hyperbaric oxygen therapy for blast-induced post-concussion syndrome and post-traumatic stress disorder. J Neurotrauma. 2012;29(1):168-85. Epub 2011/10/27. doi: 10.1089/neu.2011.1895. PubMed PMID: 22026588.

7. Morries LD, Cassano P, Henderson TA. Treatments for traumatic brain injury with emphasis on transcranial near-infrared laser phototherapy. Neuropsychiatr Dis Treat. 2015;11:2159-75. doi: 10.2147/NDT.S65809. PubMed PMID: 26347062; PMCID: PMC4550182.

8. Quirk BJ, Torbey M, Buchmann E, Verma S, Whelan HT. Near-infrared photobiomodulation in an animal model of traumatic brain injury: improvements at the behavioral and biochemical levels. Photomed Laser Surg. 2012;30(9):523-9. Epub 2012/07/17. doi: 10.1089/pho.2012.3261. PubMed PMID: 22793787; PMCID: 3423868.

9. Oron A, Oron U, Streeter J, De Taboada L, Alexandrovich A, Trembovler V, Shohami E. Near infrared transcranial laser therapy applied at various modes to mice following traumatic brain injury significantly reduces long-term neurological deficits. J Neurotrauma. 2012;29(2):401-7. Epub 2011/11/02. doi: 10.1089/neu.2011.2062. PubMed PMID: 22040267.

10. Khuman J, Zhang J, Park J, Carroll JD, Donahue C, Whalen MJ. Low-level laser light therapy improves cognitive deficits and inhibits microglial activation after controlled cortical impact in mice. J Neurotrauma. 2012;29(2):408-17. Epub 2011/08/20. doi: 10.1089/neu.2010.1745. PubMed PMID: 21851183; PMCID: 3261787.

11. Chung H, Dai T, Sharma SK, Huang YY, Carroll JD, Hamblin MR. The nuts and bolts of low-level laser (light) therapy. Ann Biomed Eng. 2012;40(2):516-33. Epub 2011/11/03. doi: 10.1007/s10439-011-0454-7. PubMed PMID: 22045511; PMCID: 3288797.

12. McCarthy TJ, De Taboada L, Hildebrandt PK, Ziemer EL, Richieri SP, Streeter J. Long-term safety of single and multiple infrared transcranial laser treatments in Sprague-Dawley rats. Photomed Laser Surg. 2010;28(5):663-7. Epub 2010/10/22. doi: 10.1089/pho.2009.2581. PubMed PMID: 20961232.

13. Naeser MA, Saltmarche A, Krengel MH, Hamblin MR, Knight JA. Improved cognitive function after transcranial, light-emitting diode treatments in chronic, traumatic brain injury: two case reports. Photomed Laser Surg. 2011;29(5):351-8. Epub 2010/12/25. doi: 10.1089/pho.2010.2814. PubMed PMID: 21182447; PMCID: 3104287.

14. Pu H, Guo Y, Zhang W, Huang L, Wang G, Liou AK, Zhang J, Zhang P, Leak RK, Wang Y, Chen J, Gao Y. Omega-3 polyunsaturated fatty acid supplementation improves neurologic recovery and attenuates white matter injury after experimental traumatic brain injury. J Cereb Blood Flow Metab. 2013;33(9): 1474-84. Epub 2013/06/27. doi: jcbfm2013108 [pii] 10.1038/jcbfm.2013.108. PubMed PMID: 23801244; PMCID: 3764381.

15. Shin SS, Dixon CE. Oral fish oil restores striatal dopamine release after traumatic brain injury. Neurosci Lett. 2011;496(3):168-71. Epub 2011/04/26. doi: S0304-3940(11)00443-5 [pii] 10.1016/j.neulet.2011.04.009. PubMed PMID: 21514362.

16. Mills JD, Bailes JE, Sedney CL, Hutchins H, Sears B. Omega-3 fatty acid supplementation and reduction of traumatic axonal injury in a rodent head injury model. J Neurosurg. 2011;114(1):77-84. Epub 2010/07/20. doi: 10.3171/2010.5.JNS08914. PubMed PMID: 20635852.

17. Bailes JE, Mills JD. Docosahexaenoic acid reduces traumatic axonal injury in a rodent head injury model. J Neurotrauma. 2010;27(9):1617-24. Epub 2010/07/06. doi: 10.1089/neu.2009.1239. PubMed PMID: 20597639.

18. Hoffmann S, Beyer C, Zendedel A. Comparative analysis of gonadal steroid-mediated neuroprotection after transient focal ischemia in rats: route of application and substrate composition. J Mol Neurosci. 2015;56(1):12-6. doi: 10.1007/s12031-014-0462-9. PubMed PMID: 25416650.

19. Hasadsri L, Wang BH, Lee JV, Erdman JW, Llano DA, Barbey AK, Wszalek T, Sharrock MF, Wang HJ. Omega-3 fatty acids as a putative treatment for traumatic brain injury. J Neurotrauma. 2013;30(11):897-906. doi: 10.1089/neu.2012.2672. PubMed PMID: 23363551.

20. Lewis M, Ghassemi P, Hibbeln J. Therapeutic use of omega-3 fatty acids in severe head trauma. Am J Emerg Med. 2013;31(1):273 e5-8. doi: 10.1016/j.ajem.2012.05.014. PubMed PMID: 22867826; PMCID: PMC3518659.

21. Scrimgeour AG, Condlin ML. Nutritional treatment for traumatic brain injury. J Neurotrauma. 2014;31(11):989-99. doi: 10.1089/neu.2013.3234. PubMed PMID: 24605947.

22. Bigford GE, Del Rossi G. Supplemental substances derived from foods as adjunctive therapeutic agents for treatment of neurodegenerative diseases and disorders. Adv Nutr. 2014;5(4):394-403. doi: 10.3945/an.113.005264. PubMed PMID: 25022989; PMCID: PMC4085188.

23. Zandi Mehran Y, Firoozabadi M, Rostami R. Improvement of neurofeedback therapy for improved attention through facilitation of brain activity using local sinusoidal extremely low frequency magnetic field exposure. Clin EEG Neurosci. 2015;46(2):100-12. doi: 10.1177/1550059414524403. PubMed PMID: 24939868.

24. Pachalska M, Lukowicz M, Kropotov JD, Herman-Sucharska I, Talar J. Evaluation of differentiated neurotherapy programs for a patient after severe TBI and long term coma using event-related potentials. Med Sci Monit. 2011;17(10):CS120-8. Epub 2011/10/01. doi: 881970 [pii]. PubMed PMID: 21959618; PMCID: 3539468.

25. Reti IM, Schwarz N, Bower A, Tibbs M, Rao V. Transcranial magnetic stimulation: A potential new treatment for depression associated with traumatic brain injury. Brain Inj. 2015:1-9. Epub 2015/05/08. doi: 10.3109/02699052.2015.1009168. PubMed PMID: 25950260.

26. Koski L, Kolivakis T, Yu C, Chen JK, Delaney S, Ptito A. Noninvasive brain stimulation for persistent postconcussion symptoms in mild traumatic brain injury. J Neurotrauma. 2015;32(1):38-44. Epub 2014/06/24. doi: 10.1089/neu.2014.3449. PubMed PMID: 24955920.

HOW YOU CAN HELP TREAT OUR VETERANS – CALL TO ACTION

Our clear **goal** for the Rocky Mountain Hyperbaric Association's *Healing Our Heroes Fund* is to help returning veterans with traumatic brain injuries and post-traumatic stress **heal**. It is our hope through the treatment provided that they can either return to the service of our country or become productive citizens once again.

In the military population, the emerging causes of TBI in veterans of Iraq and Afghanistan are blasts, blasts causing motor vehicle accidents, and gunshot wounds. Especially in exposures to blasts, unlike other causes of TBI, they are likely to produce different symptoms. For example, veterans sometimes develop post-concussive symptoms months after a blast injury. In addition, many veterans have multiple medical situations. PTSD, a history of TBI—whether mild, moderate, or acute—chronic pain, and substance abuse are all common and may complicate recovery from any diagnosis. Given these considerations, it is important to address the symptoms in a timely manner with appropriate treatment that encourages healthy behaviors.

It is through this book that we, the authors, wish to raise awareness of these terrible maladies, displaying the stark realities of what is actually happening, and establish what is being done to move ahead. The path ahead is hopeful and is being cleared by people like those mentioned in the book. It is by following their actions and helping to **spread the word** and **raise donations** that you can help a veteran.

You have read what just two Miracle Workers were able to do when others became involved and helped establish an exceptional hyperbaric program. You too can help in a variety of ways:

1. Just as the American Legion, VFW, Military Order of The Purple Heart, and generous donors have done, you can raise funds for the travel, board, and treatment of these combat veterans who can also be

healed if you donate to the Healing our Heroes Fund and send cash or a check to the Rocky Mountain Hyperbaric Association, Attention: Fund Manager Eddie Gomez at 225 South Boulder Road, #101, Louisville, CO. You will receive a tax-deductible receipt and know that you too can become a Miracle Worker.

2. Everyone knows a family or maybe a friend who may have a combat veteran who is still suffering from concussive brain damage or PTSD. Many veterans themselves still refuse to pursue HBOT therapy because they may be receiving VA disability pay or some other financial support for their wounds. You can advise them that the Rocky Mountain clinic retains all personal medical information and shares none. They may also need your help to call the clinic at 303-442-4124 and discuss how the clinic might help heal them.

3. There are also veterans who lack transportation from their homes in the Denver area to the clinic in Louisville. You or a group could offer to drive them several times a week. Usually the private home is filled with non-Denver, out-of-town patients, making it important to have local transportation. Eddie also can set up a flexible clinic schedule, which means the veteran must take forty hours of continuous treatment except on weekends. A five-day drive schedule can be developed to and from the veteran's home.

4. You may find others interested in the Miracle Workers and what they are doing. This book will be available at a much reduced price for any and all who would be interested in sharing this unique therapy program with others. Extra books are available at the clinic.

5. To date, we have very little support by our own Colorado congressmen and women. If you are in a position to influence them, we would ask that you call or write them to encourage them to visit the clinic and support this proven therapy. Oklahoma and Indiana are now developing state-sponsored HBOT therapy programs. Since

Colorado already has its own HBOT clinic that has now treated and healed more than 250 TBI-PTSD veterans, why not sponsor this program too?

Your Veteran Advocates,
Colonel Bob Fischer and Grady Birdsong

Rocky Mountain Hyperbaric Association for Brain Injuries
% Healing Our Heroes Fund
225 South Boulder Road
Suite 101
Louisville, Colorado 80027
303-442-4124 b
303-666-2112 f
Eddie@rockymountainhyperbaricassociation.org
www.rmhabi.org

EXHIBITS

Exhibit A

DoD Numbers for Traumatic Brain Injury
Worldwide-Total (All Severities) = 333,169

■ Penetrating ■ Severe ■ Moderate ■ Mild ■ Not Classifiable

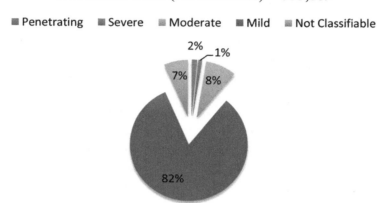

**Numbers courtesy of DoD and Defense and Veterans Brain Injury Center
for 2000-2015 Q1-Q2 Aug 18, 2015 http://dvbic.dcoe.mil/**
~ graph courtesy of Grady Birdsong

Penetrating TBI, or open head injury, is generally characterized by: A head injury in which the scalp, skull, and dura matter (the outer layer of the meninges) are penetrated.

Severe TBI is characterized by: Confused or disoriented state that lasts more than twenty-four hours; or loss of consciousness for more than twenty-four hours; or memory loss for more than seven days.

Moderate TBI is characterized by: Confused or disoriented state that lasts more than twenty-four hours; or loss of consciousness for up to thirty minutes, but less than twenty-four hours; or memory loss lasting greater than twenty-four hours but less than seven days; or meets criteria for Mild TBI.

Mild/Not Classifiable TBI is characterized by: Confused or disoriented state that lasts less than twenty-four hours; or loss of consciousness for up to thirty minutes, or memory loss lasting less than twenty-four hours.

Exhibit B

DoD Worldwide Traumatic Brain Injury in the U. S. Armed Forces
2000-2015 Q1-Q2 (as of Aug 18, 2015)

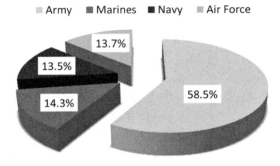

Army   Marines   Navy   Air Force

13.7%
13.5%
14.3%
58.5%

**Numbers courtesy of DoD and Defense and Veterans Brain Injury Center
for 2000-2015 Q1-Q2 Aug 18, 2015 http://dvbic.dcoe.mil/**
~ graph courtesy of Grady Birdsong

ARMY Traumatic Brain Injuries

Active	148,135
Guard	32,103
Reserve	14,552
Total	**194,790**

MARINES Traumatic Brain Injuries

Active	43,908
Reserve	3,997
Total	**47,905**

NAVY Traumatic Brain Injuries

Active	41,419
Reserve	3,465
Total	**44,884**

AIR FORCE Traumatic Brain Injuries

Active	38,744
Guard	4,317
Reserve	2,529
Total	**45,590**

Exhibit C

Computerized Neurocognitive Status Vital Signs (CNS-VS) Validity

This screening tool summarizes test results in accordance with clinical practice standards. For example, a Validity Indicator (VI) is used and alerts the clinician to the possibility of an invalid test. If there is willful exaggeration, misunderstanding, poor vision, sickness, need of sleep, etc., a retest can be done.

Without going into a long discussion of how VI is calculated, it is known that the *Clinical Domain* validity indicators are based on summary data from extensive trialing over the years. Its scoring algorithm is based on research presented (Detecting Invalidity in Neurocognitive Tests) at the International Society for CNS Clinical Trials and Methodology (ISCTM) in 2009. This is a time-tested assessment method.

Screening and Assessment Tool

CNS-VS is not intended to replace formal lengthy neuropsychological evaluation. CNS-VS is a neuropsychological testing screener and is used extensively in over 9,000 clinician sites in North America. It is widely acknowledged in peer-reviewed journals to be very suitable for use in screening and serial assessment measurement.

Patient Profile:	Percentile Range				> 74	25 - 74	9 - 24	2 - 8	< 2
	Standard Score Range				> 109	90 - 109	80 - 89	70 - 79	< 70
Domain Scores	Subject Score	Standard Score	Percentile	VI**	Above	Average	Low Average	Low	Very Low
Neurocognition Index (NCI)	NA	85	16	Yes			X		
Composite Memory	102	103	58	Yes		X			
Verbal Memory	61	93	32	Yes		X			
Visual Memory	61	110	75	Yes	X				
Psychomotor Speed	174	93	32	Yes		X			
Reaction Time*	555	107	68	Yes		X			
Complex Attention*	21	56	1	Yes					X
Cognitive Flexibility	26	63	1	Yes					X
Processing Speed	48	79	8	Yes				X	
Executive Function	34	75	5	Yes				X	
Simple Visual Attention	40	108	70	Yes		X			
Motor Speed	124	105	63	Yes		X			

Figure 3. CNS-VS report example
~ courtesy of CNS-VS www.CNSVS.com

Scoring

To give the reader an overview of this base test and how it is scored, refer to Figure 3. The above patient profile example grades a number of domains with respect to his/her brain functionality. In the *Subject Scores* column, each score is simply the number of correct responses, incorrect responses, and reaction times. The *Standard Scores* colored column records raw scoring, comparing it to other people in that age bracket. In the *Percentile Score* column, scoring is computed from an index of that age bracket's expected scores. Higher scores are better.

For simplicity in explaining this example, let's focus on the Domain Score, *Visual Memory*, which scored at 110 in the *Standard Score* column (Figure 3). This is above the expected level in the Standard Score range of >109. This is an age-matched score relative to other people in a normative sample. In the *Percentile Score* column, a score of 75 Percentile also is above average. Scoring above the average range of 25-74 (Figure 4), a 75 Percentile score is a mathematical computation of the Standard Score results and an index of how the patient scored compared to others in that age group.

Evaluation of the severity of the impairment in this testing software is based on an age-matched normative comparison database (Figure 4). This allows for a more complete assessment of the patient's condition.

Severity Classification Grade:

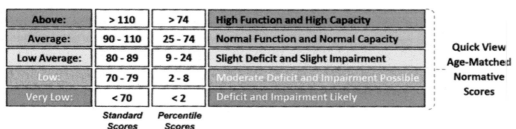

Above:	> 110	> 74	High Function and High Capacity	
Average:	90 - 110	25 - 74	Normal Function and Normal Capacity	Quick View
Low Average:	80 - 89	9 - 24	Slight Deficit and Slight Impairment	Age-Matched
Low:	70 - 79	2 - 8	Moderate Deficit and Impairment Possible	Normative
Very Low:	< 70	< 2	Deficit and Impairment Likely	Scores

Standard Scores   Percentile Scores

Figure 4.
~ courtesy of CNS –VS www.CNSVS.com

Recall Before and After Treatment

The *Visual Memory* test, to pick one of the Domains, indicates how well a person recognizes, remembers, and recalls geometric figures. The relevance of this testing is related to remembering graphic instructions, navigating, operating machines, recalling images, and remembering an abundance of events.

By utilizing this software testing capability the clinic gathers comprehensive quantitative analysis of veteran patient progress from their first HBOT treatment until completion of their therapy.

★ ★ ★ ★ ★

Exhibit D

CereScan of Littleton, Colorado an integral member of the Miracle Workers team, provides another key metric/measurement tool in the overall HBOT therapy with its SPECT brain imaging technology which clearly shows healing as a result of HBOT.

Brain SPECT & Assessment Report on Colonel George Corbin, U.S. Army (Ret)

The following has been transcribed and interpreted by Grady Birdsong from an interview with Shane Quint CTO, CereScan, and excerpts from Dr. Hipskind's diagnosis letter of Colonel Corbin. The attending and interpreting CereScan physician and former Chief Medical Advisor, S. Gregory Hipskind, M.D. and Ph.D., with permission from Colonel Corbin, shared the following assessment report, originally notated and filed on 28 August, 2014:

Clinical overview of patient

Colonel George Corbin, U. S. Army (Ret) sought hyperbaric oxygen treatment at the Rocky Mountain Hyperbaric Institute, beginning in 2014. Colonel Corbin came to the clinic with a history of significant traumatic brain injury and related post-traumatic stress. A veteran of multiple combat tours in Iraq and Afghanistan, from 2003 through 2012, he began to notice progressive cognitive, emotional, and constitutional (systemic) symptoms, starting in 2010. Perplexed, Colonel Corbin began searching for answers. Finding the Healing Our Heroes program in Louisville, Colorado, Corbin immediately took leave and came to the Rocky Mountain clinic. After the first forty hours of prescribed treatment, he came back to the clinic for a second series of forty dives and was recommended for an optional pre-treatment and follow-up brain SPECT imaging evaluation with CereScan during his second session of forty dives. Below are the results of that procedure:

Patient's self-reported symptoms

Confusion and cognitive function problems

Cognitive decline and long-term memory loss

Short-term memory problems

Difficulty integrating information

Difficulty learning new things

Difficulty performing familiar tasks

Problems with language and word finding

Disorientation to time and place

Problems with abstract thinking

Anger management problems

Impulse control problems

Difficulty with concentration

Difficulty following instructions

Problems paying attention, losing things, making careless mistakes

Distractibility and disorganization, excessive sadness

Sleeping too much and fatigue, loss of interest in things

Loss of motivation, mood swings and racing thoughts

General anxiety and panic attacks, social anxiety

Flashbacks of trauma and nightmares

Balance problems, ringing in ears

GI problems, hot flashes, and muscle spasms.

Interval history and response to treatment

Doctor Hipskind notes on the Corbin history: "Since his previous visit, George has undergone approximately 40 Hyperbaric Oxygen Treatments (HBOT) under the direction of Dr. Julie Stapleton. He reports significant improvement in his headaches, cognition, memory, affect, and sleep disturbance. He states he is much less irritable and denies any suicidal ideation. He is quite pleased with his clinical response to the HBOT therapy."

History of brain injury

Doctor Hipskind writes a brief summary of Corbin's brain injuries: "George reports that his most significant head injury happened in 2011. He was riding on rough terrain in a military Humvee. The driver hit a bump and the patient flew up out of his seat and hit his head on the steel roof of the vehicle. He recalls 'seeing stars,' but remained conscious. He suffered post-injury confusion, headaches, and floaters in his vision. He has been in and out of combat zones over 9 years with exposure to burn pits, fires, and incoming and outgoing artillery and mortars."

SPECT Brain Imaging Methods

Taken from Dr. Hipskind's analysis, the following was annotated from his summary notes prior to measuring Colonel Corbin: "During the baseline study, the patient is placed in a comfortable reclining chair and an intravenous line is started. The patient then is allowed to acclimate to a quiet semi-darkened room with sound dampening headphones in place, according to established practice guidelines. The 99mTc-labeled HMPAO tracer then is injected through the IV line and flushed with saline. The perfusion pattern of the patient's brain is captured during the subsequent 3 minutes. This is the pattern that is analyzed and interpreted. After injection, the patient remains in the quiet semi-darkened room for an additional 15 minutes. Scans are acquired 60 minutes after tracer injection."

This process, as explained by Shane Quint, Chief Technology Officer at CereScan, is to measure blood flow to the different parts of the brain using pharmaceuticals that attach to the blood and emit gamma rays, allowing image tracking of blood flow in the various parts of the brain. This is known as perfusion. The camera set up to rotate around the patient's brain picks up those gamma rays. This testing is prefaced by placing the patient in a relaxed environment.

"During the concentration task, the patient is placed in a quiet room and an IV line is started. The patient performs a Stroop colored word test on a laptop computer. Approximately 5 minutes into the computer task, the 99mTc-labeled HMPAO tracer is injected through the IV line and flushed with saline. The patient completes the computer task and 60 minutes after injection, the patient is scanned."

The Stroop test is a simple test, as explained by Quint. It flashes words and sentences up on the screen and feeds the patient a line such as, "the ball was red" but the word "red" is shown in the color green. This is to mildly confuse. Then, the patient is asked a question about what color is the word red. What that does is increase activity in the frontal lobe (blood flow), all the time being measured by the SPECT scan. If the area is damaged or has an attention deficit disorder (ADD) then the brain may not call for additional blood. This test helps to rule out ADD versus brain injury.

"Scans are obtained with a Siemens' Symbia E SPECT camera with low-energy, high resolution (LEHR) parallel hole collimation. Counts are collected in a 64 x 64 matrix with 32 stops of 6 degrees each. Total counts exceeded 5 million. Data is zoomed to 1.78, corrected for motion artifact and filtered using a Butterworth filter at .6 with an order of 6. Attenuation correction is performed. The volume is masked to exclude non-neural structures. There is no post-filtering. Data is presented in horizontal, sagittal, and frontal views with 4mm sections. Statistical parametric analysis is performed using Segami Corporation software relative to a normative database containing 64 individuals."

Quint explains the resolution of imaging in lay terms. There are 64 x 64 points of collection by another 64 slices collected. In total, a summation of 262,000 points of gamma ray collection allows the building of a three-dimensional image of the brain on a timeline.

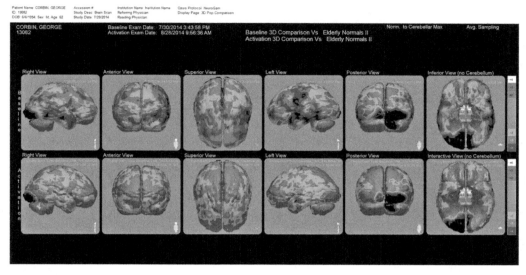

Figure 1. 3D SPECT scans Before (Top Row-labeled baseline) and After (Lower Row-labeled Activation) HBOT treatment. Gray areas indicate normal
~ courtesy of CereScan & Colonel George Corbin, USA (Ret) http://www.cerescan.com

Findings

Doctor Hipskind annotates and comments on the two methods of CereScan imaging: "A baseline follow-up study only was performed. No abnormal motion or artifact was noted. Images were compared to his previous pre-treatment brain SPECT report of 7/29/2104. Overall, his resting cortical hypo perfusion is markedly improved relative to his previous scan in all lobes of the brain to some degree. There is particular improvement in the bilateral dorsolateral frontal and anterior temporal areas of the cortex. Subcortically, there is interval perfusion improvement in both the basal ganglia and the thalami."

Shane Quint, CTO, explains what stands out on the 3D SPECT scan (Figure 1). On the right-hand side is a color chart. Those indicate standard deviation from normal. **Gray is normal and the whole gray range is normal. Light green is 2 standard deviations below normal.** The blue colors, light blue and dark blue colors, 3 and 4 standard deviations below normal, and a black ends up being 5 below normal. On the upper side of the scale +2 through +4 and so on, standard deviations above normal. In the Superior view before treatment (labeled Baseline) and after treatment (labeled Activation) you can visualize significant improvement in Colonel Corbin's condition. The left view also indicates marked improvement.

Impressions

"This remains an abnormal brain SPECT study but shows fairly dramatic interval improvement in all cortical and subcortical areas relative to the previous scan of July 29, 2014, as described above, presumably largely the result of the HBOT intervention. Correlation with all other clinical history is advised."

Additional Impressions

"None"

Sincerely,
Signed by...
S. Gregory Hipskind, MD PhD
Chief Medical Advisor, CereScan
Certified Nuclear Neurologist
Certified Brain Injury Specialist, 2010
Colorado Medical License No. DR-22702

http://www.cerescan.com/
991 Southpark Drive #200, Littleton, CO 80120
866-722-4806

ADDITIONAL RECOGNITION

Distinguished Service Award 2015

This award to Rocky Mountain Hyperbaric Institute and Rocky Mountain Hyperbaric Association for Brain Injuries for distinguished service was presented by the United Veterans Committee (UVC) of Colorado, which is a nonprofit coalition of fifty chartered and federally recognized veterans service organizations and affiliates established in 1972. The vision of UVC is to render productive advocacy for Colorado veterans and their families. The mission is to represent all associated member organizations with one centrally focused voice in Colorado, working with state legislature ensuring full support on all veteran issues and concerns, and informing/educating Coloradan's about veteran needs.

L-R: Sloan Gibson, Deputy Secretary of Veteran Affairs, Pepe Ramirez, Eddie Gomez, and Steven A. Rylant, President UVC
~ courtesy of Darrell Meyers, American Legion Observer

The award reads:

"In recognition and appreciation for your outstanding work in providing hyperbaric oxygen therapy to veterans of brain injuries, especially traumatic brain injuries, post-traumatic stress disorder and post-concussion syndrome.

You continue to educate the general public regarding the positive healing effects and benefits. Your partnership with the "Healing Our Heroes" program supports veterans by assisting with treatment payments. Your work has dramatically improved the lives of Veterans and their families."

April 19, 2015

American Legion Care Giver Award of the Year
~ courtesy of Grady Birdsong, Veteran Advocate

American Legion Care Giver Award of the Year
~ courtesy of Grady Birdsong, Veteran Advocate

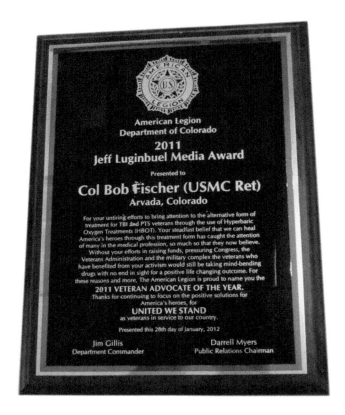

American Legion Jeff Luginbuel Media Award for Colonel Bob Fischer
~ courtesy of Grady Birdsong, Veteran Advocate

This plaque represents one of the tributes received from the
brave Australian Army soldiers who were treated and healed
of their TBI and PTSD at the Rocky Mountain Hyperbaric clinic.

Compliments of Australia's 2nd Commando Regiment
~ courtesy of Grady Birdsong, Veteran Advocate

CHAPTER ENDNOTES

Chapter 1

1. *Hyperbaric Oxygen Therapy Can Improve Post-Concussion Syndrome Years after Mild Traumatic Brain Injury – Randomized Prospective Trial*, by Rahav Boussi-Gross equal contributor, Haim Golan equal contributor, Gregori Fishlev, Yair Bechor, Olga Volkov, Jacob Bergan, Mony Friedman, Dan Hoofien, Nathan Shlamkovitch, Eshel Ben-Jacob, Shai Efrati, Published: Nov 15, 2013 (Copyright: © Boussi-Gross et al.).

2. *Hyperbaric Oxygen Induces Late Neuroplasticity in Post Stroke Patients – Randomized, Prospective Trial*, by Efrati, Shai; Fishlev, Gregori; Bechor, Yair; Volkov, Olga; Bergan, Jacob; Kliakhandler, Kostantin; Kamiager Izhak; Gal, Nachum; Friedman, Mony; Ben-Jacob, Eshel; Golan, Haim, Published: Jan 15, 2013 (Article in Academic Journal of PLOS ONE; Jan2013, Vol. 8 Issue 1).

Chapter 10

1. A Phase I Study of Low-Pressure Hyperbaric Oxygen Therapy for Blast-Induced Poet-Concussion Syndrome and Poet-Traumatic Stress Disorder, Paul G. Harch, Susan R. Andrews, Edward F. Fogarty, Daniel Amen, John C. Pezzullo, Juliette Lucarini, Claire Aubrey, Derek V. Taylor, Paul K. Staab, and Keith W. Van Meter., Journal of Neurotrauma 29: 168-185; January 1, 2012.

2. Hyperbaric Oxygen Therapy Can Improve Post Concussion Syndrome Years after Mild Traumatic Brain Injury - Randomized Prospective Trial, Rahav Boussi-Gross equal contributor, Haim Golan equal contributor, Gregori Fishlev, Yair Bechor, Olga Volkov, Jacob Bergan, Mony Friedman, Dan Hoofien, Nathan Shlamkovitch, Eshel Ben-Jacob, and Shai Efrati, Published: Nov 15, 2013 (Article in Academic Journal of PLoS ONE; Nov2013; Vol. 8 Issue 1).

Afterword

1. The Nobel Prize in Physiology or Medicine (2005) was awarded jointly to Barry J. Marshall and J. Robin Warren "for their discovery of the bacterium Helicobacter pylori and its role in gastritis and peptic ulcer disease." Both believed that peptic ulcer was caused by bacteria, but the majority of MDs and scientists believed it was due to spicy food and stress. Their groundbreaking histological work clearly demonstrated that twenty-six out of thirty stomach biopsies showed clear association with only one type of bacteria (H. pylori). Yet, the medical community was unconvinced. Dr. Marshall drank a culture of H. pylori (1984) and developed peptic ulcers. Biopsies of Dr. Marshall's stomach showed the culprit bacteria in the ulcer. He cured his peptic ulcers with antibiotics. It took well over ten years before the medical community accepted the results and started to apply them to patients.

GLOSSARY OF TERMS

Angiogenesis. Development of new blood vessels from pre-existing vessels. It occurs in the body for healing wounds and restoring blood flow in damaged tissues after injuries.

ATA. Used to express absolute pressure, which is zero-referenced against a perfect vacuum, using an absolute scale, so it is equal to gauge pressure plus atmospheric pressure. 1.5 ATA is almost equivalent to 17 feet below sea level.

CHS. Certified Hyperbaric Specialist. The highest certification level of technical expertise in hyperbaric therapy.

CHT. Certified Hyperbaric Technologist. An older certification and no longer used. See CHS.

CO. Commanding Officer.

CMS. Centers for Medicare & Medicaid Services.

CNS Vital Signs. Computerized Neurocognitive Status Vital Signs. See CNS-VS.

CNS-VS. Computerized Neurocognitive Status Vital Signs. A clinical testing procedure (software) that evaluates and summarizes the thinking, reasoning, and remembering of TBI patients in graph form. A tool used to measure the progress of TBI-PTSD patients as a result of hyperbaric treatment. www.cnsvs.com.

CTO. Chief Technology Officer.

DoD. Department of Defense.

EMDR. Eye movement desensitization and reprocessing. A psychotherapy developed by Francine Shapiro that emphasizes disturbing memories as the cause of PTSD. The goal of EMDR is to reduce the long-lasting effects of distressing memories by developing adaptive coping mechanisms.

Electromagnetic Pulse Therapy. The arena of electromagnetic (non-light based) systems is a non-invasive method used to stimulate small regions of the brain and relies on direct electrical or magnetic scalp stimulation.

FDA. Food and Drug Administration.

GI. Gastrointestinal, referring collectively to the stomach and small and large intestine areas.

Gunny. E-7 enlisted Marine referenced as a Gunnery Sergeant or "Gunny."

HBOT. Hyperbaric Oxygen Therapy.

Humvee. The High Mobility Multipurpose Wheeled Vehicle (HMMWV), commonly known as the Humvee, is a four-wheel drive military light truck. Primarily used by the United States military, it is also used by numerous other countries and organizations and even in civilian adaptations. The Humvee's widespread use in the Persian Gulf War, where it negotiated the treacherous desert sand with ease, helped inspire the civilian Hummer automotive marque.

IED. See Improvised Explosive Device.

Improvised Explosive Device. A bomb constructed and deployed in ways other than in conventional military action. It may be constructed of conventional military explosives, such as an artillery round, attached to a detonating mechanism. IEDs are commonly used as roadside bombs.

Light Therapy (Laser, LED, LLLT, IR, NIR and Blue). Emerging industry for light-based medical technologies.

MCRD. Marine Corps Recruit Depot. There are only two recruit training depots in the Marine Corps—one at Parris Island, SC, and the other in San Diego, CA.

MP. Military Police.

Neovascularization. The formation of functional microvascular networks (blood paths) in red blood cell perfusion. Neo, meaning new, and vascular, meaning of or relating to a channel for the conveyance of a body fluid (as blood of an animal or sap of a plant) or to a system of such channels.

Neuroplasticity. The ability of the brain to heal, restore and revitalize itself.

Neurotrauma. Refers to injuries to a nerve or nerves, especially part of the central nervous system (brain and spinal column).

Nutraceuticals. A portmanteau of the words "nutrition" and "pharmaceutical," coined in 1989 by Stephen L. DeFelice, founder and chairman of the Foundation of Innovation Medicine. The term is applied to products that range from isolated nutrients, dietary supplements and herbal products, to specific diets and processed foods such as cereals, soups, and beverages. Emerging research focuses on using nutrition and key dietary compounds to promote nervous system regeneration.

OEF. Operation Enduring Freedom (encompasses the Global War on Terrorism).

OIF. Operation Iraqi Freedom.

PCS. Post-Concussion Syndrome.

Physiatry. Area of medicine specializing in non-surgical physical medicine and rehabilitation.

Post-Traumatic Stress Disorder. A mental disorder that can develop after a person is exposed to a traumatic event, such as sexual assault, warfare, traffic collisions, or other threats on a person's life. Symptoms may include disturbing thoughts, feelings, or dreams related to the events, mental or physical distress to trauma-related cues, attempts to avoid trauma-related cues, alterations in how a person thinks feels, and acts, and last for periods longer than a month. Those with post-traumatic stress disorder are at a higher risk of suicide.

PTSD. See Post-Traumatic Stress Disorder.

RMHA. Rocky Mountain Hyperbaric Association for Brain Injuries.

RMHI. Rocky Mountain Hyperbaric Institute.

Rocket-Propelled Grenade. A shoulder-launched anti-tank weapon.

RPG. See Rocket-Propelled Grenade.

SFTT – Soldiers for the Truth. An organization started by Colonel David Hackworth. The mission is to provide results-oriented treatment for American Veterans affected by TBI and PTSD. The foundation has established a strategic coalition of resources to facilitate emotional, behavioral and cognitive wellness for Veterans and their families. www.sftt.org.

Signature Wounds. Both traumatic brain injury and post-traumatic stress are the signature wounds of the last wars in the Middle East from 2001 to present.

TBI. See Traumatic Brain Injury.

Traumatic Brain Injury. Brain dysfunction caused by an outside force, usually a violent blow to the head.

TreatNOW. Coalition of citizens, veterans, institutions and service organizations working to get help to our brain-injured troops and others who are suffering from concussions, traumatic brain injury (TBI), and post-traumatic stress disorder (PTSD). www.treatnow.org,

USMC. United States Marine Corps.

UVC. United Veterans Committee of Colorado. A non-profit coalition of fifty chartered and federally-recognized veterans service organizations and affiliates established in 1972.

VA. United States Department of Veterans Affairs.

501c3. The most common type of tax-exempt nonprofit organization exempt from federal income tax if its activities are for the following purposes: charitable, religious, educational, scientific, literary, testing for public safety, fostering amateur sports, etc.

ACKNOWLEDGMENTS

The Miracle Workers and their Veteran Advocates are deeply indebted to a large number of people whose selfless, generous, and professional support made it possible to write this book. Since the inception of the Rocky Mountain Hyperbaric Institute and its nonprofit Rocky Mountain Hyperbaric Association for Brain Injuries, an impressive number of technical, financial, and medical personnel, as well as true believers, have stepped up to volunteer their skills, reputations, and funds. This has resulted in the successful treatment and healing of our combat veterans who suffered signature wounds in the wars from Desert Storm to the battlefields of Iraq and Afghanistan.

The continuing and overwhelming support of so many has resulted in a proven and effective hyperbaric oxygen treatment program that has healed more than 250 veterans to date. Within that same treatment program and its innovative plan is an integrated post-traumatic stress disorder (PTSD) program, the only **integrated program** of its kind in America. Speaking for the principals in this special story, we want to recognize and honor all of those who came to assist as and when it has been most needed.

The Honorable Marty Hoffman – Former Secretary of The Army who almost daily walked the Halls of Congress to present the merits of hyperbaric therapy to those who would listen, from the FDA, DoD, and the American Legion, to a wide range of medical and professional groups.

Caleb Gates – Vice President of Denver Investments, who from his first exposure to the clinic's work, has been unwavering in his support, through both his professional contacts and the annual fundraising. His efforts continue to sustain the Healing Our Heroes nonprofit fund that heals veterans totally free of charge.

Jim Wallace – Former Chairman & CEO of Grease Monkey Intl. Inc. and Founder of BWAB Oil Exploration Corporation, partnered with Caleb Gates to raise funds for the Healing Our Heroes program. Seventy to eighty veterans per year now benefit from their generous funding alone.

Paul Harch MD – HBOT pioneer and leading U.S. therapist, who not only established the proven treatment protocol but conducted and documented his test trials of combat veterans suffering concussive brain damage. His treatment process is the standard for all HBOT clinics.

Brigadier General Stephen N. Xenakis MD – Author of *Agent Orange De-Ja Vu* that exposed the failure of the U.S. Army and DoD to provide effective treatment for combat veterans with TBI and PTSD. His timely advice and insights facilitated the integration of these two therapies at the clinic.

Xavier Figueroa PhD – Director of Clinical & Scientific Research of Scientia Incognita and co-founder of the Brain Health & Healing Foundation has authored instructive, challenging articles to support HBOT. We are honored to have him write our "Afterword" that looks at the future of hyperbaric therapy and other related treatments for TBI and post-traumatic stress.

Dr. James K. Wright, Colonel USAF (Ret) – As former Chief, HBOT Medicine Research, Davis Hyperbaric Laboratory, Brook's Air Force Base, TX, he was a pioneer in this new therapy. While stationed at Hurlburt AFB, FL, he joined Dr. Eddie Zant, a private physician, to treat, heal, and return several Airmen to active duty. He writes of this experience in our "Foreword."

T. Scott Martin, Tom Petrie and Rich Eichler – For their dedication and support of the Healing Our Heroes nonprofit fund as gala organizers, we honor their dedication to the special HBOT treatment and healing of our combat veterans with TBI and PTSD. Having raised more than one million dollars for the Rocky Mountain Hyperbaric Association, they have provided former and future hyperbaric therapy for hundreds of these veterans.

Dr. William Duncan – Vice President, Government Affairs, International Hyperbaric Medical Assn., whose dedication and persistence resulted in the Oklahoma Veteran Traumatic Brain Injury and Recovery Act that is now the law and fully supported by that state's governor and the OSU Medical School. It is the first state program of its kind to address TBI therapy.

Robert L. Beckman, PhD, Captain, USAF, 1970 - 1975, Vietnam Veteran – Close associate of Marty Hoffmann who has assumed the management, coordination, and integration of the jointly-developed TreatNOW program, whose objective, through education, is to stop the high rate of veteran suicides. He is also the major source of hyperbaric therapy information and HBOT supporting clinics.

Brigadier General James L. Bauerle, U.S. Army (Ret) – Veteran advocate leading the effort at the Military/Veterans Coalition of Indiana. They are military, veterans, and uniformed-service organizations that promote military awareness and advocacy throughout Indiana. BG Bauerle is a tireless leader, committed to the veterans, their families, and survivors in promoting efforts to support health, compensatory, entitlement, readiness, retention, and benefit matters within the state. His Indiana coalition groups have sent a number of their veterans to the Rocky Mountain Hyperbaric Institute for treatment.

American Legion Department of Indiana – For their continued support to treat Indiana veterans and for being significant contributors to the Rocky Mountain Hyperbaric Association's nonprofit fund.

Pat Smith, Adjutant, American Legion of Colorado – For his early and continuous financial and media support of the Rocky Mountain Hyperbaric Institute plus his introduction of Sergeant Margaux Mange and her amazing HBOT treatment story when little was known about the program. It was Pat and his people who led the way for HBOT to become a recognized treatment throughout the American Legion nationwide.

Darrel Myers, Editor, American Legion *Observer* – For dynamic articles that provided nationwide HBOT coverage and profiled the clinic's first veterans and their successful treatment. Darrel also initiated three Veteran Advocate and Care-Giver Awards to honor Ryan and Eddie.

Tom Bock, Past National Commander 2005-2006, American Legion – For his early support in contacting and educating nationwide America Legion chapters and assisting a local fledgling hyperbaric clinic and its nonprofit Healing Our Heroes fund that now treats combat veterans totally free.

Ralph Bozella, Chairman, American Legion's Veteran Affairs & Rehabilitation Division and former President of the United Veterans Committee – As a thirty-eight-year member of the American Legion, he testified about the VA's sad veteran TBI treatment to the House Committee on Veterans

Affairs and also got the Legion to place HBOT therapy for TBI first on its national agenda list. He also initiated the UVC's Meritorious Service Award for Ryan and Eddie.

Sandy Flint, Chairman, Colorado America Legion Wounded Warrior Program – For her immediate financial assistance to house and feed the first combat veterans treated at the clinic and then assist in the clinic's expansion and movement to its current professional site.

Staff Sergeant Dean Sanchez, USMC Wounded Warrior Representative – For his untiring, dedicated assistance to medically discharged veterans suffering TBI and PTSD, who he found, fed, and funded before he obtained their VA disability rating and other medical support.

Christine Cook, Wounded Warrior Advocate, VA Hospital, Denver, Colorado – As the mentor and advisor of the first veterans to get HBOT treatment, she also alerted the clinic that PTSD regression had occurred in some veterans. This led to the integrated PTSD program.

Robert Alvarez, Marine Veteran, Fort Carson Returning Veteran's Counseler – First Veteran Advocate, Rocky Mountain Hyperbaric Association for Brain Injuries, who recommended the first TBI-PTSD soldiers for HBOT initial treatment with Dr. Paul Harch and then with Ryan's Hyperbaric Institute.

Dan Guenther, Former Director of Research for International Learning Systems, Inc., Vietnam Veteran and Author – For his advice and contribution to the Measurement & Patient Progress Chapter that illustrates and explains the clinic's purpose, processes, and sequential steps that lead to qualifying-quantifying its HBOT therapy.

Major General Michael Edwards U.S. Air Force – Adjutant General, Colorado – For support and assistance in studying the Colorado National Guard's combat experience in Desert Storm, Iraq, and Afghanistan, as well as his own personal visits to the clinic. We also appreciate his kindness in meeting our HBOT experts and national program coordinators.

Rick Baum, Veteran Advocate – For being instrumental in all of the early fundraising, not only in the American Legion, but also with the VFW and other venues. Rick continues to support Rocky Mountain Hyperbaric Association from his new home in Indiana by sending us veterans and also by raising money from there.

Debbie Quackenbush, Colorado Founder/President American Military Family, Inc. – For her affiliation and support in sending us veterans. Her organizations' mission statement: "To honor and support all members of the United States military and their families through financial assistance, emotional support, and collaborative efforts with other nonprofit organizations and volunteers who together strive to assist those serving in the time of need."

Lawrence Quinn, Veteran Advocate – For his tireless work and support of the Healing Our Heroes Benefit ride fundraiser in May 2012.

Les Metcalf, Colorado American Legion Riders, District 5 President – For leading the early effort to organize a huge Healing Our Heroes Benefit ride in May 2012.

Terri Shelefontiuk, Veteran Advocate Extraordinaire – For being a tireless advocate whose support for TBI-PTSD veterans is recognized throughout the Denver area in the American Legion. She was instrumental, along with Rick Baum and Les Metcalf, in organizing the Healing Our Heroes Benefit ride, bringing in the bands for the fundraiser, and presenting the check to RMHBOT.

Richard Hunt, Commander MOPH Department of Florida – For continuing to send us many TBI-PTSD veterans from his organization. He served with Grady Birdsong in Vietnam, 1968.

J. B. Haskins, Chief of Staff & Adjutant MOPH Department of Florida – For being a tireless veteran advocate of the U. S. Navy working with Richard Hunt in Florida to send us their veterans. J. B. is also a Vietnam Veteran.

Beau Williams, VFW Veteran Advocate – As Post Adjutant of VFW Post 4171, Golden, Colorado, he made sure that the Healing Our Heroes benefit of 2012 had a facility and helped organize this gala affair.

John and Barb Patrick, Retired – John, a close friend of the author's was a Navy Corpsman who served in Vietnam with 1st Battalion, 4th Marines in sometimes heavy combat. John went on to complete a distinguished Nursing career in the Medical field before retiring. His faithful wife Barbara, a past editor, and John together provided the authors with helpful medical advice and copy editing.

Colorado School of Trades – Donated two pistols for the Healing Our Heroes raffle, which resulted in healthy donations to the Healing Our Heroes fund.

Cooper's Troopers – An informal association of Marine and Navy veterans whose members were the first to provide essential funds for HBOT therapy, housing, and food for the earliest veterans who came to the clinic. Several troopers donated money to treat a veteran for the entire forty-day period and a few troopers were also treated and healed by this HBOT therapy.

The Military Order of the Purple Heart, Navy Seal Danny P. Dietz, Jr. Chapter 1041, Arvada, Colorado – For raising and donating significant funds for the Rocky Mountain Hyperbaric Association Healing Our Heroes fund.

CereScan – We wish to sincerely thank the CereScan management team in Denver, Colorado, for allowing us to understand their unique technology and practices that they offer by using patient diagnosis of an actual SPECT scan of a Rocky Mountain Hyperbaric Association Healing Our Heroes patient. This company and their people are on the leading edges of understanding traumatic brain injuries.

CNS Vital Signs – A hearty thank you to the management team at CNS Vital Signs for allowing us to display their powerful neurocognitive and behavioral assessment tools and technology. They are indeed a leader in providing clinicians and researchers with leading edge capabilities. We are truly grateful for them for allowing us to talk about their software.

Kristin Strohm and Natalie Tabor – Starboard Group's premier non-profit fund-raising Galas live up to the mission: to breathe new life into dynamic non-profits like the Rocky Mountain Hyperbaric Association's Healing Our Heroes Fund. Thanks to them the fund now treats and heals hundreds of combat veterans who suffer from TBI and PTSD.

Nick Zelinger, NZ Graphics – Whose expertise and professional skill have created another iconic cover for the authors as he did with our previous books, *Fortunate Passage, Covan, and Guerrilla Grunt*. They too have benefitted from his unique and inspiring cover and layout work provided to us in *The*

Miracle Workers of South Boulder Road. We are honored to have Nick as a mentor and friend.

Donna Mazzitelli, Writer, Author, Publisher, Editor and Coach – It was a pleasure to work with this fine lady, a professional with abounding knowledge of all the right things to add to this story and making it special. Our combat veterans with TBI and PTSD could not have a more clairvoyant final voice than Donna!

ABOUT THE AUTHORS

Colonel Robert L. Fischer

Bob Fischer is a 1955 Naval Academy graduate and career Marine Corps officer who retired in 1982. He was Captain of Marines on the U.S.S. Saint Paul CA-73, the 7th Fleet Flagship, from 1961 to 1963, when he studied four guerrilla wars in Southeast Asia and obtained the Malaya Jungle School Syllabus at Johore Bahru. He used the syllabus to establish the 2nd Marine Division Counter-guerrilla Warfare Center at Camp Lejeune, North Carolina. There, 20,000 Marines, Navy Seal, and Special Forces Teams were also trained. His CIPA award-winning book *Guerrilla Grunt* documents this experience. He was also a task force advisor (Covan) for the Vietnamese Marine Corps from 1966 to 1968 and wrote his book Covan about this experience. From 1977-80, as Commander of the Defense Electronics Depot, Kettering, Ohio, his workforce set the Defense Logistics Agency's all-time performance record. For this he was awarded the Defense Superior Service Medal.

In 2010, he attended a presentation by the Rocky Mountain Hyperbaric Institute and its nonprofit Rocky Mountain Hyperbaric Association for Brain Injuries, where he learned about their recently established **Healing Our Heroes** fund. His involvement with other Marine veteran organizations motivated him to become a Veteran's Advocate for the clinic. Joining Grady

Birdsong, they filmed the first veterans who received hyperbaric oxygen therapy (HBOT) in the original Boulder clinic and raised funds by presenting the unique HBOT story to groups in the Denver area. For his efforts, he was named Colorado American Legion's Veteran Advocate of the Year in 2012.

Grady T. Birdsong

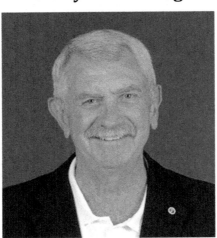

Grady T. Birdsong was raised in Kansas before enlisting in the United States Marine Corps in 1966. After serving two tours in the Northern I Corps region of Vietnam during Tet of 1968 and the DMZ in 1969, he traveled the world, enjoying a successful career in engineering, business development, marketing and technical sales in the telecommunications/data systems, information technology systems and the optical and fiber systems test industries. Additionally, Grady is the author of *A Fortunate Passage* with two EVVY awards from Colorado Independent Publishers Association (CIPA).

In 2010, Grady and Bob Fischer became Marine Corps Veteran Advocates for the Rocky Mountain Hyperbaric Institute, at a time when Ryan Fullmer and Eddie Gomez were struggling to establish their brand new HBOT clinic in the industrial area of Boulder, Colorado. It was his early filming of the clinic's first successful TBI-PTSD veteran's treatment that generated the first significant donor funds, earning $135,000. These funds enabled the HBOT clinic to move to its current site in the Professional & Medical Center in

Louisville, CO, and also to provide a nearby home for the forty-day treatment of out-of-town veterans.

Now retired, Grady lives with his wife, Pamela, in the Denver area, where he enjoys his grandchildren and spends his time writing, volunteering, and hunting big game. Grady is a graduate of Regis University in Denver, Colorado.

Both authors have an ongoing commitment to veterans of all wars and continue to advocate, inform, educate, and raise nonprofit funds. Both remain Semper Fidelis.